# ULTRALIGHTS

## The Complete Book of Flying, Training and Safety

*First flights:*

*"The first solo flight is one of the events in a pilot's life which forever remains impressed on his memory. It is a culmination of difficult hours of instruction, hard weeks of training and often years of anticipation. To be absolutely alone for the first time in the cockpit of a plane hundreds of feet above the ground is an experience never to be forgotten."*

*Charles A. Lindbergh*
*We , Grosset & Dunlap, 1927*

# ULTRALIGHTS

## The Complete Book of Flying, Training and Safety

By

Rick Carrier

*A Dolphin Book
Doubleday & Company Inc.
Garden City, New York
1985*

ONE OF THE WONDERS
IS
MAKING YOUR DREAMS
A REALITY

*Other works by Frederick (Rick) Goss Carrier*
*Books:*
>  **DIVE:** THE COMPLETE BOOK OF SKIN DIVING – WILFRED FUNK
>  **ACTION CAMERA:** SUPER EIGHT CASSETTE FILM MAKING
>  FOR THE BEGINNER – SCRIBNERS
>  **FLY:** THE COMPLETE BOOK OF SKY SAILING – McGRAW HILL

*Feature Film:*
>  **STRANGERS IN THE CITY** – Embassy Pictures

All rights reserved
Manufactured in the United States of America
First Edition

Copyright © 1985
by Frederick Goss Carrier
Art & Design

Library of Congress Cataloging
in Publication Data

Carrier, Frederick Goss, 1925-
    Ultralights: The Complete Book of Flying, Training, and Safety.

    1. Ultralight aircraft.
    2. Ultralight
TL685.15.C37    1985    629.133′343    83-40143
ISBN 0-385-19290-8

# ACKNOWLEDGEMENTS

All Photos in this book printed by MOTAL CUSTOM DARKROOMS – NEW YORK CITY.

I wish to graciously thank the following individuals, manufacturers, publications and organizations for special assistance, reference material, flight training, suggestions, permissions, photographs and the loan of equipment.

To Paul H. Poberezny, Jack Cox and Henry M. Ogrodzinski of the Experimental Aircraft Association for my invitation to the magnificent Oshkosh '83 Fly-In. For Tina Trefethen of the Eipper Aircraft Flight Team who was the pilot on my first ultralight flight in a two-place Quicksilver MX, I salute you. Also from Eipper Aircraft; Lyle Byrum, Lucky Campbell, Walter Kole, Bruce Noll and a special thank you to my flight instructor on the Quicksilver MX, Charles B. Whittelsey, II. A hug and hand shake to Pixie and Chuck Slusarczyk of CGS Hawk Aviation, Inc., who kindly introduced me to many helpful friends of ultralight flying. For Roger Worth and staff of the Cuyuna Engine Company many thanks for your patience in explaining in detail the workings of the two-cycle engine. To my friend from hang-gliding days, Tom Peghney and his co-worker at Pioneer, the builders of the Flightstar ultralight, Steve Theisen, good flying and safe landings for years to come. A smile and hale to Jo Anderson and Joe Tong, two outstanding cross country pilots.

A special *thank you* is extended to friends at American Aviation (formerly American Aerolights): Larry Newman, Brian Allen, Emory Ellis, Robert Mullikin and a warm respect for my flight instructor, Kris Williams, who gave me the opportunity to fulfill a life-long ambition to fly.

To those in communications I offer my respect and admiration for work well done. Tracy H. Knauss of *Glider Rider Magazine*; Robyn and Dave Sclair of *Ultralight Flyer Magazine*; Roy Muth of the Powered Ultralight Manufacturers Association; The AOPA Aircraft Owners and Pilots Association and their magazine, *Ultralight Pilot*; *Ultralight Aircraft*; *Air Progress ULTRALIGHTS*; *Flying*; *Sport Aviation*; *Warbirds*; and an exceptional series of articles and books on ultralights by Michael A. Markowski of Ultralight Publications that I highly recommend. Robert N. Buck's *Flying Know-How*; Sanderson's *Private Pilot Manual*; *Sky-Master International Ultralight Training System*; Van Nostrand's *Scientific Encyclopedia-Fifth Edition*; Department of Transportation-Federal Aviation Administration and all their publications relating to aviation and ultralight safety and rules and regulations.

To Myron Matzkin of Canon Camera Corporation who kindly loaned me an excellent A-1 with power-winder and a complement of sharp lenses that I used to photograph this work. To Gordon Weston and Pamela Butler who assisted me in researching the material presented here; Richard Morin. Lt. Gen. USABEC, who helped me in times of stress and to Dr. Kenneth W.M. Judy, Admiral, USABEC, who supplied the friendship and space for me to accomplish this work. To those of you I may have inadvertently omitted, my apologies and grateful thanks for all your help.

I salute you all.

From the top of the Chrysler Building in New York,
Frederick (Rick) Carrier.

# DEDICATED TO THE AMERICAN BALD EAGLE
*Our Nation's Seriously Endangered Living Symbol*

*Across the American sky
crested white clouds
float beneath the sky blue
sounds of rustling trees echo
Across the wide open quiet
of the wilderness
forever and ever*

*Bald Eagles
lovers
mated for life
fly their patterns
black and white dots swoop
beneath the sky's blue whiteness
carve invisible arcs
that only insects may notice*

*Against mountain meadows
streams of clear water bubble
Eagles soar
above the spruce
wings wide
feathers flutter
pine boughs dip
soft baby chirpings
ripple over clear lake waters
where Eagles fly to survive*

*Wings of grace and power
swoop and switch across the lake
mighty talons setting thrust
the Eagle strikes the trout
slap*

*It mantles the kill
high into the sky*

*Eagles
landing gently over their young
feed and cast food
over the crests of mountains
the fruits of this land
America*

*Bald Eagles
alive in good country
train the young
wise courage to fly
against treetopped mountains
to winter in a favorable land*

*The coolness of clouds and air
eyes muse
frosted leaves
turn my wings
against winter winds
death and life await
in the wide open quiet
of the wilderness*

*Across the American sky
with and with the wind
the Eagle flies
forever and ever and ever...*

The U.S. Bald Eagle Command is a New York based, non-profit, tax-exempt organization founded in 1975 by Frederick Goss Carrier. His experience of flying with two Bald Eagles in Colorado while researching a book on hang-gliding prompted him to create the organization, which is dedicated to saving America's most seriously endangered living symbol.

In 1982, the Bald Eagle Command was responsible for creating the National Bald Eagle Day celebrations at Independence Hall in Philadelphia. The event, highlighted by two live Bald Eagles and the original painting of the Great Seal of the United States presented to George Washington at his inauguration (1789), paid tribute to the efforts of all organizations working to preserve our national heritage and resources.

The ultimate success of Bald Eagle restoration is the reestablishment of a self-sustaining wild nesting population. Working with the U.S. Fish and Wildlife Service,* the Bald Eagle Command plans to establish a captive eagle breeding aviary and rehabilitation clinic to help repopulate America with the Bald Eagle. At present, the only major producer of Bald Eaglets from a captive breeding program is the Patuxent Wildlife Research Center, U.S. Fish and Wildlife Service, in Laurel, Maryland.

We would like your help in bringing the Bald Eagle back into the American sky.

*From 1975 to 1983, the U.S. Fish and Wildlife Service supervised translocations of 219 Bald Eagles, by *Hacking, Fostering* and *Egg Replacement*. The Hacking program was 100% successful. Six to eight week old eaglets were placed on towers located at sites where Bald Eagles can survive. The eaglets were carefully fed under controlled conditions so as not to imprint them with humans until they were capable of flight and could feed themselves. The Fostering program was only 84% successful. Two to four week old eaglets were placed in active Bald Eagle nests to be raised by foster eagle parents. Egg translocations were the least successful. Only 26% survived. Fertile "clean" eggs were used to replace infertile eggs in Bald Eagle nests where the resident eagles have a history of reproductive failure.

*Photo by William Daniel Grey, Brig. Gen. – USABEC*

OFFICERS:
Frederick Goss Carrier, President
Lynn A. Ramsey, V.P. & Secretary
Dr. Kenneth Judy, V.P. & Treasurer
H. Chester Grant, Vice-President

THE UNITED STATES OF AMERICA
**BALD EAGLE COMMAND**
SUITE 200          396 BROADWAY
NEW YORK, N.Y. 10013

A TAX-EXEMPT, NON-PROFIT ORGANIZATION

To continue the cause to save the Bald Eagle, I have given and pledged all my personal holdings, wealth and future income which I willingly contribute to the Bald Eagle charity. All income from this book goes toward the welfare and future of our nation's most seriously endangered living symbol; the Bald Eagle.

*Rick Carrier*

First Wave – D-Day – Normandy, June 6, 1944 –
Bronze Arrowhead – 5 Bronze – 1 Silver – Battle Stars.

**WARNING NOTICE**

Flying an ULTRALIGHT or any aircraft is a three-dimensional activity and as such, can lead to accidents, personal injury and possible death. To minimize your flying accident potential you should study this book along with other flying and ultralight periodicals (see listings in back of book), owners' manuals and most importantly, before you attempt to fly, you must take flight instructions and training from a Certified Flight Instructor (CFI). No book can teach you all that you need to know about flying. Remember this: Never perform aerobatics or hedge-hop.

# CONTENTS

## FLIGHT – PILOT IN COMMAND    10
SAFETY IN FLIGHT IS, IN THE FINAL ANALYSIS, THE SOLE RESPONSIBILITY OF THE PILOT-IN-COMMAND

### PREFLIGHT BRIEFING    12
HUMAN BEHAVIOR:
THE #1 CAUSE OF AVIATION ACCIDENTS

### AIRCRAFT    14
Lighter than Air: Balloons, Blimp, Dirigibles, Gliders, Hang Gliders, Sailplanes - Rotorcraft: Gyroscopes, Helicopters - Airplanes: Monoplanes, Single Engine, Multi-Engine, Amphibians, Mid-Wing, Low Wing, High Wing, Biplane, Triplane

### ULTRALIGHT AIRCRAFT STRUCTURAL COMPONENTS    16
Ultralight - Styles

### ULTRALIGHT PROPULSION SYSTEM – ENGINES    18
2-Cycle - 4-Cycle Engines

### PROPELLERS –    20
Function - Dynamics

### FEDERAL AVIATION AGENCY – RULES AND REGULATIONS FOR ULTRALIGHTS – PART 103    21

### MY FIRST FLIGHT IN AN ULTRALIGHT    31

## FLIGHT TRAINING EAGLE XL    33
YOUR FIRST STEP TO FLIGHT IN AN ULTRALIGHT

### FIRST DAY
EAGLE XL SINGLE-PLACE AIRCRAFT
American Aerolights Inc. Manufacturer - Firm Personnel and History - Quality Control - Dealer Flight Training Procedures - Eagle XL Schematic View and Specifications - Aircraft System Function - Descriptions

### PILOT BRIEFING –    44
Aerodynamics - Physical Laws Governing Flight Forces - Airfoils: Bernoulli's Principle, Newton's Laws - Gravity - Thrust - Lift - Drag - Wing-Structure Components - Relative Wind

### FLIGHT SIMULATOR – UNPOWERED    50
Pilot Experience Restrained Flight Using Aircraft Controls - Canard - Spoilerons - Wingtip Rudders - Aircraft Flight Characteristics - Takeoffs - Level Flight - Banks: Left and Right - Flares and Touchdowns - Three-Axis-Control - Pilot Preparation

### TAXI POWERED CONTROLS    56
Develop Preflight Discipline - Engine Start, Stop - Propeller Function - Taxi Techniques - Airport Rules and Right-of-Way Procedures

### TOW-TRAINING    60
Develop flying skills without the use of engine power:
Runway - Nosewheel - Steering

| **PILOT BRIEFING** | **65** |

**AIRPORT - AIRPARK RULES AND RIGHT-OF-WAY PROCEDURES**

Coronado Airport - Compass Rose - Sectional Charts - U.S. Airspace Control Zone System - Airport Traffic Patterns - Aircraft Priorities - Flying Hazards

## SECOND DAY 72

Preflight Techniques - Engine Start-Stop - Taxi - Tow-Training, Low Level - Takeoff - Landing

## THIRD DAY 78

Preflight Techniques - Engine Start-Stop - Taxi - Tow-Training: Medium Level - Takeoff - Rudder Control - Landing.

| **PILOT BRIEFING** | **82** |

Meteorology - Earth Atmosphere - Pressure - Barometers - Winds - Air Mass - Circulation - Fronts - Air Flow - Clouds - Weather - Jet Stream - Wind Sheer

## FOURTH DAY 96

Preflight Techniques - Engine Start-Stop - Taxi - Tow-Training: Medium to High Levels - Takeoff - Turns - Flares - Landings

| **PILOT BRIEFING** | **100** |

Stability - Pilot Induced Oscillation - Flight Control - Takeoff - Straight & Level Flying - Turns - Wake Turbulence

## FIFTH DAY - SOLO 104

Preflight - Engine Start-Stop - Taxi - Tow-Training: High Level - Engine Start - Takeoff - Patterns - Landings

| **PILOT BRIEFING** | **116** |

Ultralight Flight Log Records - Time - Dates - Numbers - Phonetic Alphabet - Radio Communications

# FLIGHT TRAINING QUICKSILVER MX

## FIRST DAY 118

Eipper Aircraft Inc. Manufacturer — Firm Personnel and History - Quality Control - Dealer Flight Training Procedures - Quicksilver MX Schematic View and Specifications - Aircraft System Function - Descriptions

**AIRCRAFT PREFLIGHT TECHNIQUE**
Develop consistent, accurate procedures

**TWO-PLACE MX WITH CERTIFIED FLIGHT INSTRUCTOR**
Introduction - Engine Start - Takeoff - Climbs - Turns - Descents - Stalls - Traffic Patterns - Landings

## SECOND DAY 130

Preflight - Steep Turns - Patterns - Landings - Emergency Landings

| **PILOT BRIEFING** | **132** |

Landing Procedures - Active Runways - Runway Prespective

**DOWNEY CA POLICE ULTRALIGHT** 134

## THIRD & FOURTH DAY

Preflight - Touch & Go - Landings - Turns - Patterns - Emergency Landings

## FIFTH DAY - SOLO

Preflight - Engine Start - Takeoff - Patterns - Banks - Turns - Stalls - Landing

| **ULTRALIGHT MANUFACTURERS AND AIRCRAFT NAMES** | **138** |
| **COMMENTS – STATISTICS** | **140** |

# FLIGHT
## PILOT IN COMMAND

You lift from the earth. Hundreds of pounds of thrust from a whirling fan launches you into the sky. You are a bird. Your mind and body controls the fabric, metal and power of your man-made wings.

The earth below recedes. Familiar objects on the ground shrink and take on the appearance of toys under a Christmas tree. You are isolated. Alone, high-up in a sea of invisible air that cushions and supports your temporary moments of flight.

The freedom and joy of flying with the eagles chills and thrills the senses. Yet you know an undeniable fact. Gravity never relaxes its grip. Sooner or later you must return to the earth. How and in what condition you return to the planet's surface is your responsibility. You are the pilot in command.

The moment you choose to fly you are joining the ranks of the brave. It is how you enter the realm of flight that determines your future. Your skill as a flyer predicts whether you live to see tomorrow and fly another day or if your spirit is destined to drift in space forever.

To fly an ultralight does not require great strength. You need not be a powerhouse or pump iron to achieve the rewards of personal flight. It takes a gentle touch and smooth handling of the controls along with a thorough knowledge of the forces that make flight possible or impossible to achieve mastery of the sky.

To earn your wings you must begin with a ground training program that is dedicated to perfection. The student pilot must have: a love of flight, enthusiasm, curiosity, humility, patience, prudence, good habits, abstinence from drugs and alcohol, responsibility, respect of your ultralight, boldness and a willingness to learn and never be careless. These are just a few of the determining factors that directly relate to your future as a pilot in command.

With care and a meticulous preflight attention to details the well-built ultralight can fly for many years. Airplanes (and ultralights are airplanes) are not made to think. Thinking is up to the pilot in command.

My purpose in this work is to provide you with a clear picture of what happens in an Ultralight Flight Training Program. During my training I discovered one vital fact. You can never know or experience enough or gain all of the flight experiences in any training program. A new adventure happens every time you fly into the sky.

The old excitement returns when you advance the throttle and feel the ultralight surge ahead on the beginning of the takeoff roll. The pulse quickens at the takeoff rotation and climbout. The human system begins to calm when you reach altitude and roundout into cruise. Unless hazardous weather, an engine malfunction or aircraft structural failure occurs or you get lost or run out of fuel, the rest of the flight is routine.

The mellow feelings you experience when making coordinated turns and level flight shift to high concentration when approaching your landing destination.

Your eyes scan the sky around you, looking for problems like bad weather or other aircraft in the vicinity. Ground obstacles are observed and considered along with surface wind direction and possible turbulent air on or near the runway.

When all your decisions are made and you decide to land you make the required 45° entry into the Downwind Leg of the airport traffic pattern. Your eyes become very sharp now. You are pin-pointing the exact landing spot on the runway that you are flying parallel to.

Halfway into the Downwind Leg you retard the throttle to lessen thrust and bleed-off altitude. Your airspeed is 35 MPH. A scan of the altimeter indicates 300 feet above ground level (AGL).

When the landing spot you have chosen is 45° over your shoulder, looking towards the runway, you turn the ultralight 90° into the Base Leg of the landing pattern. You are now flying perpendicular to the runway. Altitude – 250 AGL. Airspeed – 32 MPH. Intense concentration dominates all of your attention.

Anything that could interfere with your landing is seriously considered. Decisions are made. Whether to go in and land or power-up and go around for another try. Any mistake here and the purchase of the farm looms ahead.

All is clear. You turn hard 90° and enter the Final Approach Leg. Your hand grips the throttle. Gently you ease it back to idle. Eyes scan. Airspeed. 30 MPH. Center of the runway is straight ahead. Corrections by using the spoilerons level the wings with the horizon. The rudder corrects for crosswinds as the eyes fix on the landing spot you picked on the Downwind Leg. You rotate the control stick forward and angle the nose of the ultralight right at the surface of the runway. The glide angle increases sharply as the end of the runway rushes up at you. The pulse quickens but your mind instructs your body to remain loose and calm—ready to react positively and instantly to any emergency.

Seconds become critical moments in time. You are committed to land. Runway dashes zip underneath in a blur as you hold the nose down until a crash seems inevitable. At exactly the right moment you rotate the control stick back towards your belly. You are about four feet off the runway as the nose pitches up to level flight and the ultralight skims along a few feet above the surface of the earth.

Gently, delicately, you ease the stick back by millimeters as you cut the throttle to idle. At the exact moment before the wheels touch you smoothly rotate the stick all the way back. The nose pitches up. The wings catch the wind and the ultralight flares as the main wheels touch the runway. A gentle rotation of the stick forward and the nose wheel touches and you are down on the deck.

The heart slowly returns to normal as you compete the landing roll-out to a stop. Up-throttle now to taxi off the flight line to the parking area. A deep breath clears the $CO_2$ buildup as you hit the kill switch and the engine stops. You are surrounded with a quietness that seems soft compared to the roar of the engine while flying. The last piece of business of the post-flight ritual is performed. You tie-down the ultralight or dismantle it or park it in a hangar and the flight is over. As a final part of your flight discipline, you write all of the details of the flight in your log book before calling it a day.

Flight training books by themselves are no substitute for on-the-spot flight instruction from a qualified F.A.A. Certified Flight Instructor (CFI). Only the instructor is capable of observing and evaluating your progress and your abilities as a flyer. Only he or she can coach you on eliminating whatever bad habits you may have so your flights can be an exciting and rewarding experience.

Remember, flying in the atmosphere is a three dimensional event and as such presents risk of injury or death. Flying, like driving a car, motorcycle, diving or other active sports, can lead to accidents. The wise student pilot recognizes this and seeks out the best CFI instructor available as a teacher.

This book recounts exactly the training sequences I went through to learn and achieve personal flight. The events shown here happened to me. I pass this information along to you so you can see the degrees of discipline required to achieve the status of the pilot in command.

Once you solo, the next step is to develop your flight proficiency and knowledge. That task is up to you. I end these thoughts on safety with a quote from Charles A. Lindbergh's book *WE*.

*PARACHUTES:*

*"THERE IS A SAYING IN THE SERVICE ABOUT THE PARACHUTE: 'IF YOU NEED IT AND HAVEN'T GOT IT, YOU'LL NEVER NEED IT AGAIN!' THAT JUST ABOUT SUMS UP ITS VALUE TO AVIATION."*
*CHARLES A. LINDBERGH*

Good luck. Good flights and gentle landings.

F.G.C.
Pilot In Command

## PREFLIGHT BRIEFINGS

The briefings within this book are presented to familiarize you with aircraft. Ultralights are aircraft, not flying lawn chairs as they are sometimes called. Responsible, conscientious ultralight manufacturers and their flight training programs are as carefully controlled, with an emphasis on safety, as those who build airplanes and train students within the general aviation community. Not all ultralight manufacturers follow the patterns of safety and quality control. Those of you who desire to learn to fly and possibly own an ultralight are advised to use these briefings as a beginning to your training. Additional research, reading, experience and inquiry are essential to your becoming a well rounded pilot. The more you know about all the aspects of the aircraft and a high degree of enthusiasm for flying, the safer pilot you will become.

# PREFLIGHT BRIEFING
## HUMAN BEHAVIOR:
### THE #1 CAUSE OF AVIATION ACCIDENTS

Most aircraft accidents are highly preventable. Many of them have one factor in common: They are precipitated by some human failing rather than mechanical malfunction. Pilots who lived through accidents generally knew what *had gone wrong*. They were often aware of the hazards at the time they elected the "wrong" course of action, but in the interest of expediency, cost saving, self-gratification, or similar irrelevant factors, the wrong course of action was nevertheless selected.

It is a well established fact that our emotional makeup is largely responsible for the accidents we get into. Few of us are mentally ill, but not many of us are perfectly balanced either. The following list was assembled as a result of an international study on accident proneness. (Accident Proneness by Shaw and Sichel; published by Pergamon 1971.) If a person fits several of the following, he or she is likely to be accident prone.

## THE BAD ACCIDENT RISK

### Definitely Abnormal

The mentally defective or psychotic person.

The person who is extremely unintelligent, unobservant, and unadaptable.

The disorganized, disorientated, or badly disturbed person.

The badly integrated or maladjusted person.

The person with a distorted apperception of life and a distorted sense of values.

The person who is emotionally unstable and extremistic.

The person who lacks control and particularly the person who exhibits uncontrolled aggression.

The person with pronounced anti-social attitudes or criminal tendencies.

---

FEW PEOPLE BELONG IN THE GROUP ABOVE BUT IF YOU KEEP READING, YOU WILL PROBABLY RECOGNIZE SOMEONE YOU KNOW.

---

### Traits Frequently Found Among People Considered Quite Normal Are:

The selfish, self-centered, or id-directed person.

The highly competitive person.

The *over-confident, self-assertive* person.

The irritable and cantankerous person.

The person who harbors grudges, grievances, and resentment.

The blame-avoidant person who is always *ready with excuses*.

The *intolerant* and *impatient* person.

The person with marked antagonism to and *resistance against authority*.

The frustrated and discontented person.

The inadequate person with a driving *need to prove his self*.

The extremely anxious, tension-ridden, and panicky person.

The person who is unduly *sensitive to criticism*.

The helpless and inadequate person who is constantly in need of guidance and support.

The chronically indecisive person.

The person who has difficulty in concentrating.

The person who is easily influenced or intimidated.

The *careless* and *frivolous* person.

The people who are very lacking in personal insight and an appreciation of their own limitations.

The people who have the sort of personality pattern that predisposes them to *drink* or *drugs*.

The person who already gives evidence of addiction to *alcohol* or *drugs*.

The person who has suicidal tendencies or who indulge in suicide fantasies.

The people who exhibit the personality characteristics commonly associated with immaturity, such as: foolhardy impetuosity, *irresponsibility*, exhibitionism, inability to appreciate the consequences of their actions, hypersensitivity, easily aroused emotionalism, unrealistic goals, and a general *lack of self-discipline*, personal insight, worldly wisdom and common sense.

---

A LOOK AT THE ITALICIZED WORDS REVEALS MANY KEY CAUSES OF FATAL AIRCRAFT ACCIDENTS.

WHEN A TRAGIC ACCIDENT OCCURS, PEOPLE ANGRILY ASK WHY "SOMEONE" DOESN'T DO SOMETHING ABOUT THE OBVIOUS RISKS CERTAIN PILOTS TAKE?

---

### Well, "Someone" Has!

There are some *"do's and don'ts"* available to pilots that, by all statistical probabilities, could ensure the prevention of most accidents. On top of this list are the Federal Aviation Regulations and supporting Advisory Circulars. Born of a lot of know-how and practical experience, they are your *checklist for survival*. In addition, there are owner's manuals, the Airman's Information Manual, charts, operating limitations, Airworthiness Directives, and many other sources of safe operating procedures published by the Federal Aviation Administration and aircraft manufacturers. All this information serves only safety. Not to follow them is like going against your own doctor's or lawyer's advice.

### So Why Do We Still Break Rules

It is mostly for immediate gratification of some emotional need, as the above list of bad risks so clearly points out. It is common knowledge that a lot of things we often indulge in are not good for us (like smoking, speeding, over-eating, gambling, etc.). We know this with our intellect but, unfortunately, our lives are too often guided by our emotions and this certainly holds true in aviation. The existing rules would go a long way to remedy this situation. The same personality traits that cause irrational breach of safety also make a person prone to disregard the rules that would ensure safe aircraft operation.

The study also came up with a model of a good accident risk. A look at this list could almost induce one to seek a whole new philosophy of life. According to this study, when you behave as a bad accident risk, you are showing your emotional weaknesses to everyone around you. The "good accident risk" model, however, portrays an entirely different person.

## THE GOOD ACCIDENT RISK

### Traits Found In People Considered to Be Good Accident Risks Are:

The well balanced person.

The mature person.

The well-controlled person.

The person with a healthy and realistic outlook.

The person with satisfactory interpersonal relations.

The person with kindly and tolerant attitudes toward others.

The person with a well developed social and civic conscience.

The person with an ingrained sense of responsibility.

The people who are essentially moderate individuals, able to exercise adequate control over their impulses and emotions.

The people with positive attitudes who are able to assess a situation as a whole and make decisions, provided they are not too aggressive.

The contented people who are not outstanding, but who are friendly, cheerful, adaptable and accepting— provided they are reasonably intelligent, realistic, and mature.

The people who have weaknesses and limitations, but are realistically aware of them and are careful, cautious, and moderate in their behavior according to their limitations.

### The Decision Is Yours

You need not be a genius to be safe. You merely have to be an emotionally stable individual and accept the notion that you are not in possession of all the facts for all situations and be willing to accept the recommendations of those who specialize in evaluating, assessing and administering aviation procedures. One can always argue for different ways of doing things. In a large aviation community such as ours, consensus would be extremely difficult to reach. Our rules and procedures are designed to serve most of the people most of the time. A mature person can accept this gracefully and follow the rules for the benefit of all. The immature, emotionally immoderate person has to satisfy personal needs regardless of the consequences. The student carrying passengers, the non-instrument rated pilot flying into weather conditions, the local pilot buzzing the neighborhood are prime causes of accidents which often kill innocent bystanders.

It is easier to develop flying skills than good judgment. Good judgment may mean not flying when the weather is marginal (even if it is legal), or grounding yourself after taking "over the counter" medication like antihistamines or getting recurrent training after a long layoff. These are some of the decisions pilots must often make. Unfortunately, many pilots fail to make the proper decisions. This is due partly to lack of knowledge and partly to human tendency to rationalize things until they look justifiable to us. In simple terms, when we really want to do something we can generally make ourselves believe it is all right to do it.

We can't make rules for every situation. Some of the decision making is up to you. But you *can* decide on your personal limitations if you do it at a time when you are not involved in flying. Limitations to consider are fuel reserves, weather, drinking, fatigue and others. Write them down. When you have a flight decision to make, re-read them and see whether you have enough character to stick with what you decided when you were conservative and not under emotional pressure to do something foolish.

The most important decision for you to make is to stick with the published rules, procedures and recommendations. They are there for well-proven reasons and can take most hazards out of your flying. If you don't believe that, then you are really kidding yourself!

As a pilot, you hold human lives in your hands. You have a moral responsibility to operate in the safest way. If you are a bad accident risk, society will be better off if you didn't fly at all.

Reprinted from:

U.S. Department of Transportation, **Federal Aviation Administration**, Washington, D.C.

# AIRCRAFT

AIRCRAFT—Any vehicle that sustains itself in the air without attachments to the Earth is called an Aircraft, and are classified in four categories: LIGHTER THAN AIR, GLIDERS, ROTORCRAFT, AIRPLANES.

LIGHTER THAN AIR vehicles are: BALLOONS, BLIMPS and DIRIGIBLES. They are called Lighter Than Air because they use *Gas* to provide *Lift* in the form of buoyancy. Engine driven propellers provide the propulsive force for Lighter Than Air vehicles.

BALLOONS—Buoyancy by gas or hot air. Balloons have no metal framework and are classified by the FAA as ultralights. Propulsion by wind only, no engines.

**BALLOON**

BLIMP—Buoyancy by gas only. Blimps have no metal framework.

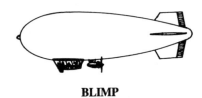
**BLIMP**

DIRIGIBLE—Buoyancy by gas. Dirigibles have a metal framework to support the fabric covering that gives the dirigible its distinctive appearance. Zeppelins are Dirigibles.

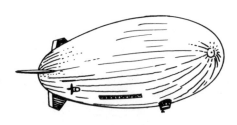
**DIRIGIBLE**

GLIDERS are aircraft that sustain flight without the use of engines. They rely on wings with a high aspect ratio and low wing loadings to achieve flight by the use of rising air thermals to keep them aloft.

**EARLY GLIDER DESIGNS**

**OCTAVE CHANUTE—1895**

**OTTO LILIENTHAL—1889**

**WRIGHT BROTHERS—1903**

HANG GLIDERS come in a variety of wing planforms (wing shape). Their propulsive force comes from the energy of the pilot who launches the Hang Glider by foot from high elevations such as mountain tops. Hang Gliders are the forerunners of Ultralight Aircraft.

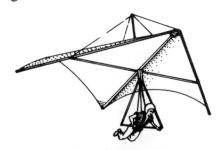
**HANG GLIDERS—1970**
Francis M. Rogallo Design

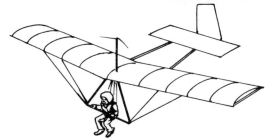

**QUICKSILVER—Frank Lovejoy Design**

**BI-PLANE HANG GLIDERS**

SAILPLANES have glide ratios of 30-to-1 or more and are known as high performance aircraft.

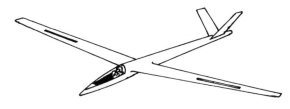

## ROTORCRAFT

GYROSCOPES are aircraft that have a free-wheeling wing rotor that produces lift. Thrust is produced by engines that are classified as tractor (engine and propeller in front) or pusher (engine and propeller in the rear). The rotor has no power.

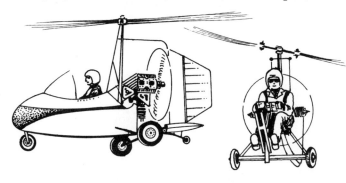

HELICOPTERS are Rotorcraft that have powered main and tail rotors that provide lift, directional stability and control.

## AIRPLANES

AIRPLANES are classified according to their intended use, power system, wing location, design and type of landing gear.

MONOPLANE—One Wing

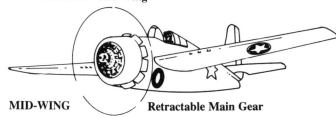

MID-WING    Retractable Main Gear

SINGLE ENGINE

LOW-WING    Tricycle-Landing Gear

MULTI ENGINE-AMPHIBIAN

HIGH-WING    Floats—Retractable Gear.

BIPLANE—Two Wings

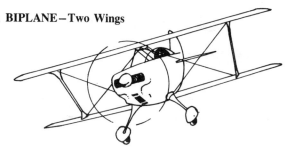

Springsteel Shock Strut Gear

TRIPLANE—Three Wings

Fixed Landing Gear—Tail Dragger

# ULTRALIGHT AIRCRAFT STRUCTURAL COMPONENTS

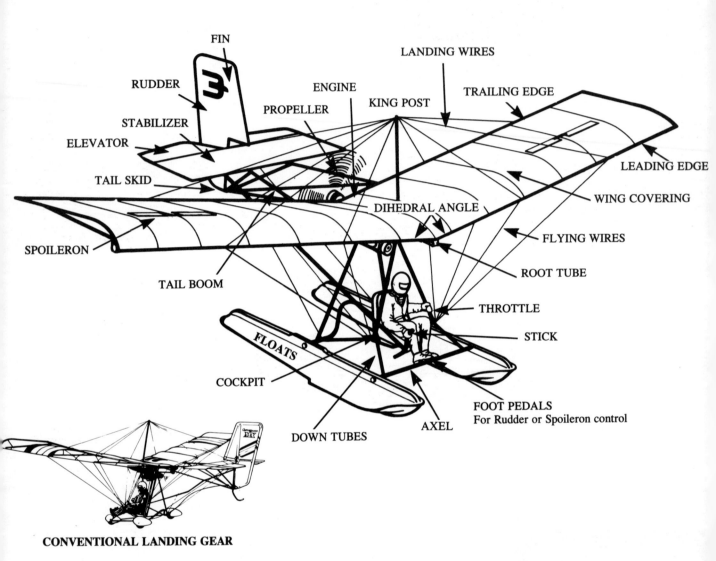

**CONVENTIONAL LANDING GEAR**

Reliably engineered, built and tested ultralight aircraft incorporate all the physical elements needed to achieve and sustain safe, controllable and stable flight.

The ultralight flight elements are: The WING (mono- or bi-plane), the FUSELAGE, an open or enclosed metal cage or pod that serves as an assembly point for: WING ATTACHMENTS, PILOT COCKPIT (controls and instruments), PROPULSION SYSTEM and LANDING GEAR (wheels, floats or snow skis).

On many ultralights the fuselage extends aft in what is called the EMPENNAGE, a cross (cruciform) comprised of either an upright or inverted FIN and movable RUDDER and a STABILIZER with a movable ELEVATOR.

The tail (Empennage) of an airplane can also have a STABILATOR which is a pivoted, horizontal one-piece elevator that rotates up and down to control the pitch of the airplane.

Airplanes can be either *pusher* (engine in rear) or *tractor* (engine in front).

# ULTRALIGHT STYLES

**LAZAIR**
High wing—strut braced twin-engine monoplane. Tractor engine.
Ultralight Sales, Ltd.

**SIROCCO**
High wing—cantilevered monoplane. Pusher engine.
Aviasud Engineering

**KASPERWING**
High wing—cable braced. Monoplane. Pusher engine.
Cascade Ultralights

**CGS HAWK**
High wing—strut braced monoplane. Pusher engine.
CGS Aviation

**FALCON**
High wing—strut braced monoplane. Pusher engine.
American Aviation

**FLIGHTSTAR**
High-wing—strut braced monoplane. Tractor engine.
Pioneer International Aircraft

**INVADER MKIII-B**
Shoulder wing—cantilevered monoplane. Pusher engine.
Ultra Efficient Products

**PARAPLANE**
Powered parachute. Pusher engine.
Paraplane Corp.

**A-10 SILVER EAGLE**
Flying wing monoplane. Pusher engine.
Mitchell Aircraft

**RALLY 2-B**
High wing—cable braced monoplane. Pusher engine.
Rotec Engineering

**JENNY**
Bi-plane—strut and cable braced. Tractor engine.
Cloud Dancer Aeroplane Works

**QUICKSILVER MX**
Float equipped. Highwing—cable braced monoplane. Pusher engine.
EIPPER AIRCRAFT

# ULTRALIGHT PROPULSION SYSTEM ENGINES

**CUYUNA TWO-STROKE ENGINE**

Ultralights are powered by a variety of reciprocating (pistons up and down) engines that began their careers as the power plants for other uses: lawn mowers, go-carts, chainsaws, snowmobiles, generators, etc. Ultralight engines are generally classified as: TWO-STROKE and FOUR-STROKE.

The dominant engine powering ultralights is the two-stroke engine. It was chosen because the two-stroker is lighter and least complex of all the other models. The two-stroke engine has disadvantages such as high revolutions per minute (6,000 RPM) to develop maximum crankshaft torque and they consume more fuel than the four-stroke version. The trade-off is less weight and complexity of construction, repair and operation.

For this briefing the Cuyuna UL-02 is used as an example.

The two-stroke engine unlike the four-stroker, does not have cylinder head intake and exhaust valves, lifters, timing gears or chains, camshafts or oil pumps.

The only moving parts of the two-stroker are the crankshaft, connecting rod and piston that serves as the opening and closing valve for fuel intake, transfer, power and exhaust each time the piston ascends and descends in the cylinder. It combines two-stroke functions each (180°) revolution of the crankshaft, whereas the four-stroke engine requires one crankshaft (180°) revolution for each of the four functions: intake, compression, power, and exhaust.

## TWO-STROKE ENGINE FUNCTION

INTAKE CYCLE—As the engine cranks over, either by an electric or pull starter, the piston ascends and descends in the cylinder 180°. As the piston ascends, the skirt of the piston passes the fuel intake port that is located near the crankcase housing. As soon as the intake port is open, a partial vacuum is created within the crankcase housing by the ascending piston. Fuel, mixed with air in the carburetor is sucked into the crankcase through the intake port. The oil-gas-air mixture circulates freely within the crankcase where the oil in the fuel lubricates the crankshaft and connecting rod wrist-pin bearings. During this first turnover of the engine there is no fuel in the cylinder head to ignite when the spark plug fires.

CYLINDER HEAD CHARGE CYCLE—When the piston starts *down* after reaching Top Dead Center (TDC) of its upward stroke, the skirt of the piston closes the fuel intake port. As the piston continues on down, the fuel trapped in the crankcase is compressed. The moment the piston head clears the fuel transfer port on the cylinder wall, the compressed fuel in the crankcase is forced up through the fuel transfer duct and into the cylinder head combustion chamber.

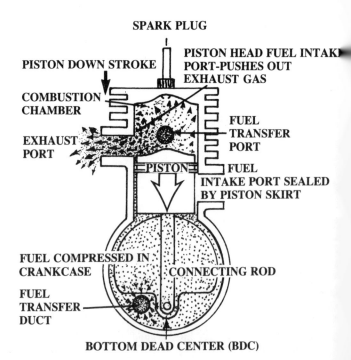

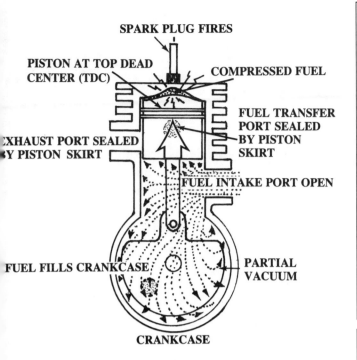

COMPRESSION CYCLE—When the crankshaft has rotated 180° from TDC to Bottom Dead Center (BDC) and the piston begins its upward stroke, it passes the exhaust and fuel transfer ports. The piston then acts as a valve and closes these two ports. The gas-oil-air mixture in the cylinder head combustion chamber is now being compressed. During this upward stroke, the moment the piston skirt passes the fuel intake port, a new fresh supply of fuel is drawn into the crankcase by the partial vacuum created by the rising piston.

IGNITION—POWER CYCLE—When the piston reaches TDC of its upward stroke—180° from BDC, the spark plug ignites the compressed fuel in the cylinder head. The explosion forces the piston down and in the process exerts torque on the crankshaft which rotates the propeller.

The moment the piston passes the exhaust and fuel transfer ports, the exhaust blows out the exhaust port and the fresh fuel being compressed in the sealed crankcase by the descending piston is forced up through the transfer duct and into the cylinder head. This rush of fresh fuel entering the cylinder head, pushes or *Loop Scavenges* the remaining residue of exhaust gasses out the exhaust port.

The propeller, flywheel, kinetic energy of the moving parts and additional cylinders keep these two cycles running smoothly at about 6,000 RPM at full throttle until the engine runs out of fuel, a malfunction occurs, or when the ignition switch is killed.

ENGINE INSTRUMENTS—The basic instruments every pilot should know and use are: The *Compass*—to determine exact headings; *Airspeed Indicators* (ASI)—to know exact airspeeds; *Tachometer*—to know the exact revolutions per minute (RPM) the engine is turning; *Cylinder Head Temperature* (CHT)—to prevent engine overheating; *Exhaust Gas Temperature* (EGT)—to tune the engine's fuel mixture—rich or lean; *Chronograph*—to compute engine time, distance traveled, fuel consumption and log records. Other available instruments are *Turn, Bank* and *Yaw* indicators. *Fuel Gauges*—thermometers to determine density, altitude. and a *Variometer* to show whether you are ascending or descending.

TWO-CYCLE ENGINE FUEL—An oil-gasoline mixture is used for efficient engine operation. The gasoline used can be premium or regular leaded or un-leaded, with an octane rating of 85 or more. **Aviation gas (AVGAS) must never be used as it can cause fouling of the carburator.** The oil used is: two-cycle, B.I.A., TC-W rated. Oil-to-gas mixture percentages are: 20-to-1 for break-in and 40-to-1 for normal operations. These mixtures must be exact. Too little or too much oil can foul the spark plugs, promote carbon build-up or freeze the engine from lack of lubrication. Clean fuel containers and filter systems must be always maintained. Dirt assures engine failure.

ENGINES are not MOTORS. Motors are exclusively powered by electricity while Engines use a variety of propulsive forces to create mechanical torque or thrust.

Engines are classified as: Reciprocating—cylinder, piston, connecting-rod and crankshaft. Turbine—rotative; Jet—thrust produced by ejection of a stream of gasses (Newton's third law of action-reaction). Rocket-gas ejection.

RECIPROCATING, TURBINES and JET ENGINES use petroleum derived fuels for combustion while ROCKETS use liquid or solid chemicals to create thrust. Steam is also used to power engines but not for airplanes.

# AIRCRAFT PROPELLER

PROPELLERS are driven by the engine or turbine either by a direct connection to the crankshaft or by a reduction drive system. Each propeller blade is an *airfoil wing* that is twisted so the propeller blade will stab or bite into the air as it rotates. The propeller provides the forward thrust necessary to achieve flight.

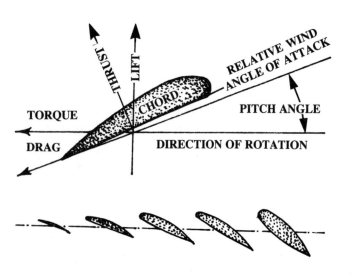

**PROPELLER AIRFOIL SECTION**

AIRFOIL — A teardrop-shaped section (chord) of a wing, aileron, stabilizer and propeller, designed to produce lift (L) as it moves through an airstream known as the Relative Wind.

The propeller is a rotating wing composed of airfoil elements, formed in a twisted shape, that rotate in circular arcs about the centerline axis of the engine drive shaft and moves forward at the same rate of speed as the airplane.

To create the greatest propeller Lift to Drag ratio (L/D), the angle of attack of each propeller airfoil element diminishes from the hub (thick airfoil) to the propeller tip (thin airfoil). This gradual decrease in airfoil angle gives the propeller its twist.

Airflow over the propeller blade airfoil creates the lift necessary to overcome drag. This lift force in propeller terminology is called Thrust.

THRUST in a propeller is determined by the total engine drive shaft power (torque) needed to rotate the propeller at maximum design efficiency.

ULTRALIGHTS use Fixed Pitch Propellers made of wood or plastic fibers.

PROPELLER TIP VELOCITIES that approach the speed of sound reduce thrust efficiency. Ultralight reduction drive systems on the engine keep tip velocities at a rotational speed between 800 and 1,000 feet per second.

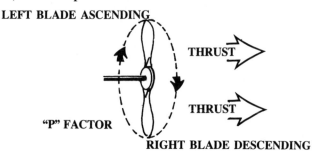

P FACTOR forces occur at high angles of attack. On a clockwise turning propeller, the descending right blade has a high angle of attack which produces more lift than the ascending propeller blade. This force differential causes the airplane to yaw to the left. This effect is known as the P-Factor and is not present when both propeller blades are at the same angle of attack as in level flight.

GYROSCOPIC PRECESSION is the result of the spinning propeller when the aircraft is in a nose high attitude. The gyroscopic effect occurs 90° opposite to the rotational direction of the propeller centerline on the engine drive shaft. This force causes the airplane to slightly yaw to the opposite direction of the propeller rotation.

Propellers, like your teeth, must be kept immaculate, free of nicks, splits, abrasions and in perfect working order. Otherwise, the propeller can and most probably will fail. The vibrations from a broken or out-of-balance prop can destroy an ultralight when under full rotational power.

# FAA RULES AND REGULATIONS FOR ULTRALIGHT VEHICLES—PART 103

**Subpart A—General**

Sec.
103.1 Applicability.
103.3 Inspection requirements.
103.5 Waivers.
103.7 Certification and registration.

**Subpart B—Operating Rules**

103.9 Hazardous operations.
103.11 Daylight operations.
103.13 Operations near aircraft, right-of-way rules.
103.15 Operations over congested areas.
103.17 Operations in certain airspace.
103.19 Operations in prohibited or restricted areas.
103.21 Visual reference to the surface.
103.23 Flight visibility and cloud clearance requirements.

Authority: Secs. 307, 313(a), 601(a), 602, and 603, Federal Aviation Act of 1958 (49 U.C.S. 1348, 1354(a), 1421(a), 1422, and 1423); sec. 6(c), Department of Transportation Act (49 U.S.C. 1655(c).

---

## Subpart A—General

### § 103.1 Applicability.

This part prescribes rules governing the operation of ultralight vehicles in the United States. For the purposes of this part, an ultralight vehicle is a vehicle that:

(a) Is used or intended to be used for manned operation in the air by a single occupant;

(b) Is used or intended to be used for recreation or sport purposes only;

(c) Does not have any U.S. or foreign airworthiness certificate; and

(d) If unpowered, weighs less than 155 pounds; or

(e) If powered:

(1) Weighs less than 254 pounds empty weight, excluding floats and safety devices which are intended for deployment in a potentially catastrophic situation;

(2) Has a fuel capacity not exceeding 5 U.S. gallons;

(3) Is not capable of more than 55 knots calibrated airspeed at full power in level flight; and

(4) Has a power-off stall speed which does not exceed 24 knots calibrated airspeed.

### § 103.3 Inspection requirements.

(a) Any person operating an ultralight vehicle under this part shall, upon request, allow the Administrator, or his designee, to inspect the vehicle to determine the applicability of this part.

(b) The pilot or operator of an ultralight vehicle must, upon request of the Administrator, furnish satisfactory evidence that the vehicle is subject only to the provisions of this part.

### § 103.5 Waivers.

No person may conduct operations that require a deviation from this part except under a written waiver issued by the Administrator.

### § 103.7 Certification and registration.

(a) Notwithstanding any other section pertaining to certification of aircraft or their parts or equipment, ultralight vehicles and their component parts and equipment are not required to meet the airworthiness certification standards specified for aircraft or to have certificates of airworthiness.

(b) Notwithstanding any other section pertaining to airman certification, operators of ultralight vehicles are not required to meet any aeronautical knowledge, age, or experience requirements to operate those vehicles or to have airman or medical certificates.

(c) Notwithstanding any other section pertaining to registration and marking of aircraft, ultralight vehicles are not required to be registered or to bear markings of any type.

## Subpart B—Operating Rules

### § 103.9 Hazardous operations.

(a) No person may operate any ultralight vehicle in a manner that creates a hazard to other persons or property.

(b) No person may allow an object to be dropped from an ultralight vehicle if such action creates a hazard to other persons or property.

### § 103.11 Daylight operations.

(a) No person may operate an ultralight vehicle except between the hours of sunrise and sunset.

(b) Notwithstanding paragraph (a) of this section, ultralight vehicles may be operated during the twilight periods 30 minutes before official sunrise and 30 minutes after official sunset or, in Alaska, during the period of civil twilight as defined in the Air Almanac, if:

(1) The vehicle is equipped with an operating anticollision light visible for at least 3 statute miles; and

(2) All operations are conducted in uncontrolled airspace.

### § 103.13 Operation near aircraft; Right-of-way rules.

(a) Each person operating an ultralight vehicle shall maintain vigilance so as to see and avoid aircraft and shall yield the right-of-way to all aircraft.

(b) No person may operate an ultralight vehicle in a manner that creates a collision hazard with respect to any aircraft.

(c) Powered ultralights shall yield the right-of-way to unpowered ultralights.

### § 103.15 Operations over congested areas.

No person may operate an ultralight vehicle over any congested area of a city, town, or settlement, or over any open air assembly of persons.

### § 103.17 Operations in certain airspace.

No person may operate an ultralight vehicle within an airport traffic area, control zone, terminal control area, or positive control area unless that person has prior authorization from the air traffic control facility having jurisdiction over that airspace.

### § 103.19 Operations in prohibited or restricted areas.

No person may operate an ultralight vehicle in prohibited or restricted areas unless that person has permission from the using or controlling agency, as appropriate.

### § 103.21 Visual reference with the surface.

No person may operate an ultralight vehicle except by visual reference with the surface.

### § 103.23 Flight visibility and cloud clearance requirements.

No person may operate an ultralight vehicle when the flight visibility or distance from clouds is less than that in the table, as appropriate:

The following is a condensed version of F.A.A. Advisory Circular 103-7 (issued 1/30/84) which defines ULTRALIGHT vehicles and the proper way to assure that FAR-103 applies to a specific ultralight.

### APPLICABILITY OF PART 103.

a. *Probably the single most critical determination* which must be made is whether or not your vehicle and the operations you have planned are permitted under Part 103. *The fact that you are operating a vehicle which is called or advertised as a "powered ultralight," "hang glider," or "hang balloon" is not an assurance that it can be operated as an ultralight vehicle under part 103.* There are a number of elements contained in § 103.1 which make up the definition of the "ultralight vehicle." If you fail to meet any one of the elements, you may not operate under Part 103. Any operations conducted without meeting *all* of the elements are subject to all aircraft certification, pilot certification, equipment requirements, and aircraft operating rules applicable to the particular operation.

b. *The FAA realizes that it is possible to design an ultralight* which, on paper, meets the requirements of § 103.1, but in reality does not. However, the designers, manufacturers of the kits, and builders are not responsible to the FAA for meeting those requirements. *Operators of ultralights should bear in mind that they are responsible for meeting § 103.1 during each flight. The FAA will hold the operator of a given flight responsible if it is later determined that the ultralight did not meet the applicability for operations under Part 103.* Be wary of any designs which are advertised as meeting the requirements for use as an ultralight vehicle, yet provide for performance or other design innovations which are not in concert with any element of § 103.1. The FAA may inspect any ultralight which appears, by design or performance, to not comply with § 103.1.

c. *If the FAA Determines Your Ultralight Was Not Eligible for Operation as an Ultralight Vehicle.* If your ultralight does not meet § 103.1, it must be operated in accordance with applicable aircraft regulations. You will be subject to enforcement action ($1000 civil penalty for each violation) for each operation of that aircraft.

### ELEMENTS MAKING UP THE DEFINITION OF AN ULTRALIGHT VEHICLE.

a. *Single Occupant Only.* Any provision for more than one occupant automatically disqualifies any ultralight under Part 103. (Two-place ultralights – use by qualified trainers are the only exemptions.)

b. *Sport or Recreational Use Only.* NO aerial advertising – towing banners, loudspeakers, programmed lights, smoke writing, dropping leaflets, advertising on wings, crop dusting, surveying, patrolling, carrying freight for hire, pilot compensation for hire or given a discount on the price of an ultralight to perform a task for the seller.

*Exemptions* – Renting an ultralight is permitted. Prizes for competition in recognition of pilot performance, writing books about ultralights, discounts on ultralight prices. Air shows and Events – providing you do not benefit directly from the proceeds as the organizer or producer of the event.

c. *No Airworthiness Certificate.* You cannot fly an ultralight under FAR-103 and as a certified aircraft interchangeably. To fly under FAR-103 you must turn in to the issuing authority any airworthiness certificates. To operate an ultralight as a certified aircraft, the pilot must have the proper certification. Make and Model ultralights may be certified for one use and also flown under FAR-103 providing the unlicensed ultralight pilot turns in any airworthiness certificates before flying the aircraft.

A Technical Standards Committee approved by the FAA will issue documents to ultralight owners that give the owner's name and address, the ultralight's make and model number, fuel capacity, speed, safety devices, performance and weight. This document will prove to any F.A.A. inspector that may question your ultralight applicability to FAR-103. For the complete text of F.A.A. Advisory Circular 103-7, write: U.S. Department of Transportation, Federal Aviation Administration, 800 Independence Ave., S.W., Washington, DC 20591.

### ULTRALIGHT DATA PLATE

The EAA Ultralight Vehicle Data Plate shown above when appropriately engraved shows that the vehicle complies with FAR Part 103. The Plate, which is very much like the Data Plates used in the amateur-built aircraft movement, is available free of charge to all EAA members, EAA Ultralight Association members, and at the nominal fee of $3.00 to non-members.

The EAA Ultralight Assn. is located at Whitman Airfield, Oshkosh, Wisconsin 54903-2591.

All serious-minded ultralight enthusiasts should become members of the EAA. Their publications alone are worth joining.

# ULTRALIGHT VEHICLE OPERATIONS—AIRPORTS, AIR TRAFFIC CONTROL, AND WEATHER   FAA ADVISORY CIRCULAR (6/23/83):

The sport of hang gliding has advanced dramatically since the Federal Aviation Administration (FAA) first issued Advisory Circular No. 60-10, "Recommended Safety Parameters for the Operation of Hang Gliders," on May 16, 1974. The purpose of that advisory circular was to provide guidance to the hang gliding community without the need for Federal regulation. The response to the guidelines of the advisory circular was excellent, and for the period immediately following its issuance many of its safety goals were maintained. But, as the sport advanced, the performance capabilities and popularity of these vehicles increased. Many unpowered gliders became capable of soaring to altitudes more than 10,000 feet above the launch point, and flight distances could exceed 100 miles. The addition of powerplants and controllable aerodynamic surfaces created vehicles which approximate the operational capabilities of fixed-wing aircraft. And with the greatly increased number of these vehicles, the operation of ultralight vehicles became a significant factor in aviation safety.

On October 4, 1982, a new Federal Aviation Regulation, Part 103, became effective and provided for the safe integration of ultralight vehicle operations into the National Airspace System. In conjunction with Part 103, the ultralight community is being encouraged to adopt good operating practices. This advisory circular is intended to assist the ultralight operator in attaining that goal.

Comments and questions concerning information contained in this advisory circular should be directed to Federal Aviation Administration, Airspace and Air Traffic Rules Branch (AAT-230), 800 Independence Avenue S.W., Washington, D.C. 20591.

## AIRPORTS AND ULTRALIGHT FLIGHTPARKS

**WHERE TO TAKE OFF AND LAND.** One of the questions most frequently asked by the ultralight pilot is, "Where can I safely and legally take off and land my ultralight?" The following information is designed to assist the ultralight pilot in understanding the different types of operations, both on and off airport, and the recommended procedures for obtaining permission to operate ultralight vehicles.

**Existing Airports.** Currently, there are approximately 16,000 public use and private airports and seaplane bases in the United states. The vast majority of these facilities may be suitable and compatible for safe ultralight operations. Information on their location may be obtained from various sources, such as FAA publications (i.e., Airport/Facility Directory, aeronautical charts, etc.) which may be purchased at most local airports. Also, ultralight user organizations have comprehensive airport listings which usually include a description of the facility.

Some of these airports have their air traffic directly controlled by an air traffic control tower. Use of these airports requires prior permission of airport management and the local air traffic control authority (see FAR Part 103.17). Since the volume of aircraft operating at these airports is usually significantly higher, ultralight operators may find operations at these airports to be less desirable than operations at uncontrolled airports.

There are many airports where air traffic is not controlled by an air traffic control tower and the traffic activity level is usually low. These airports are referred to as "uncontrolled airports." Use of these airports by ultralight vehicles may require prior permission of the airport operator. When seeking access to these airports, ultralight operators should remember that even though the airport may be tax supported, airport management has the responsibility for determining the compatibility of operating the various classes of aircraft on the airport. If an ultralight can be safely operated at the airport, then permission to operate the ultralight vehicle may be granted. Safety of aircraft operations on the airport is always the prime consideration.

**Abandoned Airports.** Since 1970, approximately 3,000 airports have been abandoned because of a lack of activity, financial problems, or other related reasons. The majority of these airports are located in rural areas, privately owned, and possibly well-suited for ultralight training and other activities. Many state aeronautical organizations have knowledge of recently abandoned facilities and should be able to assist you in finding these sites. It may be possible to obtain permission of the property owner to reactivate certain of these facilities for ultralight operations.

**Open Space Operating Areas.** One of the prime advantages of ultralight operation is the vehicle's ability to operate in small areas. FAR Part 103 does not prohibit ultralight take-off and landing from open areas, providing the operation does not overfly congested areas. Good judgement still dictates that an ultralight pilot obtain prior permission from the landowner and be familiar with the terrain and obstructions at any location where operations are intended. For the operation of hang gliders, special consideration should be given to the terrain surrounding the launch site. In many cases these terrain features will influence the ability of the unpowered craft to return to the launch site.

## OPERATION OF A FLIGHTPARK.

Anyone wishing to establish a site for the operation of ultralight vehicles should be aware of the following Federal, state, and local regulatory requirements which may apply to these operations:

**Federal Requirements.** Unless the site is to be used solely in VFR weather conditions for a period of less than 30 consecutive days with not more than 10 operations per day during this period, notification of the intent to establish a flight park is required under the provisions of FAR Part 157, Notification of Construction, Alteration, Activation, and Deactivation of Airports. FAA Form 7480-1, which is used to provide this notice (as well as guidance in its preparation) is available from any FAA

Division or Airports District/Field Office. The FAA uses the information provided in the notice to advise on the effect of the establishment of the site on the use of navigable airspace by aircraft. Advisory Circular 70-2, Airspace Utilization Considerations in the Proposed Construction, Alteration, Activation and Deactivation of Airports, describes some of the factors which affect airspace utilization. Failure to provide the required notice violates Section 901 of the Federal Aviation Act of 1958 and carries a possible civil penalty.

**State Requirements.** Many state aviation departments require approval and a license for the establishment of a site for aeronautical operations. The potential ultralight flightpark developer should contact the state aviation authorities to determine state requirements.

**Local Requirements.** Most communities have established zoning laws, building codes, fire regulations, and other legal requirements to provide for the safety and comfort of the citizenry. A thorough study of these requirements should be made to determine their effect on the establishment and operation of an ultralight flightpark.

## STANDARDS FOR THE FLIGHTPARK LAYOUT.
The FAA has no standards for the geometric design of an airport built to exclusively serve ultralight vehicles. However, several ultralight organizations provide information which may be useful for the establishment of an ultralight flightpark as a separate entity. FAA Advisory Circular 150/5300-4B, Utility Airports—Air Access to National Transportation, intended for airports serving aircraft with approach speeds less than 121 knots, provides guidance which may also be helpful in developing an operational site for ultralight aircraft.

**Noise Considerations.** Perhaps the most limiting factor in the operation of ultralights is the noise emitted from the vehicle. Unless proper measures are taken in the design and operation of ultralights, public annoyance to the noise may result in restrictive local and state regulations. Acceptance by the public of recreational sport flying is significantly tied to the potential for annoyance from the vehicle's noise.

Significant progress has been made by ultralight manufacturers to quiet engine, exhaust, and propeller noises. As these systems continue to improve, so will the acceptance of the ultralight vehicle. However, these improvements are only half of the story. Ultralight operation in a manner sensitive to the possible annoyance of those on the ground is the other. It is probably the most important factor in gaining acceptance by the general public.

Airport owners/operators have been trying for years to establish operations compatible with the needs of adjacent communities. The acceptance of ultralight operations by a community will depend in a large part on its perception of how additional operations by ultralights will affect the airport's overall compatibility with its neighbors. Careful planning by ultralight operators in integrating their vehicles into the existing operation, will go a long way in making acceptance a reality.

The FAA has begun ultralight noise testing. Preliminary results indicate that, in absolute noise levels, the ultralight is no louder at 1,000 feet AGL than some popular two seat single engine aircraft. The slower speed of the ultralight does result in longer periods of exposure to noise and is a significant factor in the annoyance perceived from such overflight. Another consideration is the lower altitude at which many ultralight operations take place. This causes an increase in the intensity of sound during fly-over and is a significant factor in determining the annoyance caused by noise.

FAR Part 103 prohibits operations of ultralights over congested areas. Ultralight pilots should be aware that, while their vehicles may not be operating directly over congested areas, their vehicles' noise may carry to the residents of a nearby congested area.

## FLIGHTPARK DATA.
Once the ultralight flightpark is activated by the operator and the FAA is notified, an Airport Master Record (FAA Form 5010-2) is prepared by the FAA. This is a computerized record of data describing the flightpark's facilities and services. Each year, a copy of this Airport Master Record is mailed to the flightpark operator with a request to verify and update the data. The information collected by the FAA is available upon request to Government agencies, aviation organizations, aviation industries, and private individuals. Future informational needs for ultralight flightpark directories, charting, etc., can be supplied from computerized data summaries derived from the Airport Master Record.

## AIR TRAFFIC CONTROL AND RADIO COMMUNICATIONS

**GENERAL.** The rapid growth and popularity of ultralight vehicles and the increased number of operations require the highest degree of vigilance on the part of ultralight operators to see-and-avoid other ultralight vehicles and aircraft. Some of these operations involve authorization from air traffic control. The purpose of this chapter is to assist the ultralight operator in understanding the airspace, operations with air traffic control, and the use of radio communications.

**AIR TRAFFIC CONTROL (ATC) AND AIRSPACE.** Even though ultralight vehicle operators are not required to demonstrate any aeronautical knowledge or experience requirements, failure to recognize and avoid certain airspace can be hazardous and may be in violation of the Federal Aviation Regulations. FAR 103.17 states that no person may operate an ultralight vehicle within an Airport Traffic Area, Control Zone, Terminal Control Area or Positive Control Area unless that person has prior authorization from the air traffic control facility having jurisdiction over the airspace. The airspace areas requiring ATC authorization that you, as an ultralight operator, are most likely to come in contact with are the Airport Traffic Area, Control Zone and Terminal Control Area.

**AIRSPACE AREAS.**

**What is an Airport Traffic Area (ATA)?** An Airport Traffic Area is airspace within a radius of 5 statute miles from the center of an airport, with an operating control tower, that extends upward from the surface to, but not including, an altitude 3,000 feet above the elevation of an airport. For the purpose of ultralight operations, flight within the ATA requires specific authorization from the air traffic control tower. Although most ATA's are not depicted on charts, any airport symbol on the sectional chart that is blue in color indicates the presence of an air traffic control tower. During the time that tower is in operation, an ATA exists (see **Airspace and the Chart** (Page 66))

**What is a Control Zone?** A Control Zone may include one or more airports and is normally a circular area within a radius of 5 statute miles around an airport. The vertical limits of a control zone begin at the surface and extend upward to 14,500 feet mean sea level (MSL). Some control zones have rectangular extensions to include the arrival and departure paths for pilots operating primarily with reference to their aircraft instruments. The entire area of a control zone is considered controlled airspace, but not all airports have a control zone. Where a control zone exists, it is depicted on sectional charts by the use of dashed lines. For the purpose of ultralight operations, flight within the control zone requires authorization from the air traffic facility controlling that area. (Page 66-68)

**What is a Terminal Control Area (TCA)?** At the present time there are 23 Terminal Control Areas. TCA's are in place around many of the high density airports in the country. They extend upward from the surface in the center and usually have multiple rings of airspace which extend outward horizontally. Its appearance closely resembles an inverted wedding cake, with both lower and upper limits for each ring. The presence of a TCA is characterized on a sectional chart by blue outlines of the TCA limits around a major airport. All operations within the rings of a TCA requires authorization from air traffic control (see **Airspace** (Page 66-68))

**What is Positive Control Area (PCA)?** Positive Control Area is the area which overlies the continental United States at 18,000 feet and above. All operations conducted in PCA are done so with the authority of air traffic control. Aircraft operating at these higher altitudes are required to carry additional radio equipment and their pilots must be rated for instrument flight. Although ultralights are not faced with specific equipment requirements for entry into PCA, ATC authorization is required. Requests for such flights will be thoroughly reviewed prior to any decision to authorize operations in PCA by an ultralight.

**How Do I Get ATC Authorization?** Requests for authorization to operate an ultralight vehicle into one of the above named areas should be made by writing, telephoning, or visiting the air traffic control facility having jurisdiction over the airspace in which you wish to operate. Requests for such authorization via air traffic control radio communication frequencies will normally not be accepted, since it may interfere with the separation of aircraft.

**What is Uncontrolled Airspace?** Uncontrolled airspace is the area in which air traffic control separation services are not provided. This area is usually below 1,200 feet above ground level (AGL). When nearing airports with established instrument approaches, the ceiling of uncontrolled airspace usually lowers to 700 feet AGL, and, if a control zone exists, uncontrolled airspace remains outside of the control zone horizontal limits, thus putting the airport within controlled airspace. In some geographic areas, primarily west of the Mississippi River, uncontrolled airspace ceilings are above 1,200 feet AGL. This is an exception, rather than the rule. The ceiling of uncontrolled airspace may be determined by reference to Sectional Aeronautical Charts used for aviation.

**What is Controlled Airspace?** Controlled airspace is the area in which air traffic control separation services are available for aircraft. The base of controlling airspace usually begins at 1,200 feet AGL and extends upward. When nearing airports with established instrument approaches the base of controlled airspace usually lowers to 700 feet AGL, and, if a control zone exists, the base of controlled airspace begins at the surface within the horizontal limits of the control zone.

**Airspace and the Chart.** Sectional Aeronautical Charts, often called "sectionals", are published by the National Oceanic and Atmospheric Administration (NOAA) and are revised on a semi-annual basis. Sectionals depict information for the use of pilots who are operating with visual reference to the earth's surface. Each sectional has a legend printed on its endflap. Of particular interest to the ultralight operator, is the portion entitled "Airport Traffic Service and Airspace Information." This portion of the legend gives information which will enable you to locate the floor of controlled airspace, prohibited and restricted areas, TCA's, control zones, tower controlled airports, obstructions, and other useful information. Sectional charts may be purchased from local airport operators, user organizations, and directly from the NOAA, Washington, D.C. Assistance in learning how to use sectional charts should be readily available from any FAA-certificated flight or ground instructor.

**Special Military Activity.** There are special routes, known as Military Training Routes (MTR's), which have been developed across the country for military training in "low level" combat tactics. Generally, MTR's are established below 10,000 feet MSL for operations are speeds in excess of 250 knots and will include operations by both fighter and cargo type aircraft. The routes at 1,500 feet AGL and below are developed primarily to be flown in visual flight weather conditions. The sectional charts depict regularly established MTR's as shaded gray lines with an associated visual rules (VR) or instrument rules (IR) numbered identifier. Nonparticipating flights are not prohibited from flying within an MTR, but extra caution to see-and-avoid these operations is imperative in attaining the greatest practical level of safety. Ultralight pilots and flightpark operators should contact the nearest Flight Service Station (FSS) to obtain information on the route usage in their vicinity. Information available includes times of scheduled activity, altitudes in use, and actual route width. Route width varies for each MTR and can extend several miles on either side of the line depicted on sectional charts.

Also, throughout the year, the military conducts special operations which may be held on a one-time basis in a specific geographical location. Information pertaining to such operations is usually available through the FSS system. When requesting MTR and special activity information, ultralight operators should give the FSS their area of intended operation and permit the FSS specialist to identify the MTR routes and special activities which could be a factor. Information on FSS's may be found on page 29.

**TRAFFIC PATTERNS AND OPERATIONS IN THE VICINITY OF AN AIRPORT.** Since the speed and operating characteristics of an ultralight vehicle may be incompatible with many aircraft, it is essential that you stay alert by looking for and avoiding other traffic. Be especially aware of the possibility that a faster craft might overtake your ultralight. Ultralight operators should be especially vigilant for aircraft operating around an airport. Traffic pattern altitudes for propeller driven aircraft generally extend from 600 to 1500 feet above the ground and aircraft are often at these altitudes within 5 miles of the airport. Also, because of the possible effects of wake turbulence, operations in close proximity to aircraft of greater speed and weight should be avoided.

Preparatory to landing at an uncontrolled airport, the pilot should be concerned with landing direction indications on the airport. Such indicators include wind socks, wind tees, tetrahedrons, traffic pattern indicators, and the direction of other fixed-wing operations.

Wind socks operate freely and are subject to the forces of wind for direction. Wind tees may move freely or be aligned manually indicating the preferred landing direction. A tetrahedron is a large kite-shaped indicator sometimes located beside the runway and may move freely or be set manually. The small end of the tetrahedron points in the preferred direction of landing.

Many airports have standardized traffic patterns which rely on all turns in the pattern being made to the left. Traffic pattern indicators are used when there is a variation from the normal left traffic pattern. They are located either in a segmented circle with the wind sock or tetrahedron, or may be located near the end of the applicable runway. If the pilot will mentally enlarge the indicator for the runway to be used, the direction of turns will become readily apparent. Airports which have parallel runways may have both left and right traffic patterns operating at the same time.

Also, some airports may have a specific area designated for ultralight operations. Look for any indications that landings are to be made on other than the main runway and adjust your flight path so as to not conflict with operations to the main runway.

Regardless of wind indicators or traffic patterns, it is wise to scan the airport surface and the surrounding airspace for flights that may be operating in a different manner. The governing factor as to which runway is in use is the direction and strength of the wind. It is the responsibility of pilots to determine the safe landing direction for their craft. The indicators are there to assist you in operating safely, but they are not meant to be a substitute for careful vigilance and good judgement.

(See Page 69)

**OPERATIONS AT AIRPORTS WITH A CONTROL TOWER.** If you are operating into or out of an airport with a control tower expect to be segregated from all nonultralight aircraft in the traffic pattern, in the use of runways, and on the airport surface. Please take special notice of the word "segregate." FAA air traffic controllers have been advised to authorize ultralight operations only if they will not interfere with and can be kept relatively clear of normal aircraft operations. Certificated aircraft receive separation services. These will not be available to ultralight pilots. Rather, ultralight pilots will be expected to separate themselves from each other and also to remain clear of all normal aircraft operations. When requesting to operate at a tower controlled airport, or within the airport traffic area, expect the controllers to provide you instructions as to what areas to avoid. These instructions may include route and altitude information as well as a specified landing area. Specific times during which to operate may also be authorized. For operators equipped with two-way radios, see paragraph below. It is important that ultralight operators understand the responsibility for avoiding a conflict with aircraft and other ultralights is theirs, and theirs alone.

**USE OF A TWO-WAY RADIO.** The following information provides guidelines for the use of a two-way radio while operating an ultralight.

**Communications with Air Traffic Control.** In all radio communications with air traffic control, ultralight operators should state the word "ultralight" followed by the call letters assigned by the F.C.C. on your radio license, i.e., "Ultralight 12593U." Use of the following radio communication practices will result in the controller having a better understanding of your request and enhance the safety of your flight.

(1) Determine the correct frequency from a Sectional Aeronautical Chart.

(2) Contact the air traffic control tower prior to entering the area for which you are requesting authorization.

(3) Speak slowly and distinctly. If you do not get an immediate reply, wait a few moments, then repeat your request. The controller may be busy and you may not be hearing all of the transmissions the controller is hearing.

(4) State the facility you are calling, your ultralight identification, altitude, and location relative to the airport. Example: "Sample Tower, Ultralight 12593U Six Miles Southwest at 1,000 feet." If you are on the ground at the airport, give your position on the airport.

(5) Wait for the tower to respond before stating any further information.

(6) Once two-way communications are established, briefly state your request.

(7) Keep in mind at all times your responsibility to remain clear of all other aircraft and ultralights. Further, remember your responsibility to remain clear of any area for which an

authorization is required, but has not been received.

(8) On occasion, air traffic control will deny authorization to operate in a specific area. This is not unique to ultralights. At times, certificated pilots in sophisticated aircraft are also denied access to certain areas. Factors affecting authorization are the nature of the requested operation, the effect on other operations that may already be taking place, controlled workload, and equipment or facility limitations. The ultimate reason remains the same . . . SAFETY.

**Communications at Uncontrolled Airports.** An uncontrolled airport is an airport where the control tower or where the control tower is not currently in operation. This does not mean that two-way communications are not used. Quite the contrary. A considerable amount of useful information is passed back and forth among pilots and the operators of airport advisory frequencies. Information such as runway in use, surface winds, other aircraft known to be in the area, and any unusual activities, such as parachuting, may be available.

There are three primary ways for ultralight operators, who are radio equipped, to communicate their intentions and obtain airport/traffic information when operating at a landing area that does not have an operating control tower:

(1) by communicating with an FAA flight service station located on the airport;

(2) by communicating with a local airport advisory operator located at the airport; or

(3) by making self-announced broadcasts of intentions over a commonly used frequency for operations at that airport.

The key to communicating at uncontrolled airports is selection of the correct Common Traffic Advisory Frequency (CTAF). A more detailed explanation of CTAF and traffic advisory practices and good operating procedures can be found in FAA Advisory Circular 90-42C and the Airman's Information Manual. Additionally, the Airport/Facility Directory provides information on which frequency to use at a particular airport.

**Traffic Advisory Practices at Uncontrolled Airports.** In all radio communications, ultralight operators should state the word "ultralight" followed by the call letters assigned by the F.C.C. on your radio license, i.e., "Ultralight 12593U."

(1) Select the correct frequency, many of which can be found on Sectional Aeronautical Charts.

(2) Contact the airport advisory service prior to entering the area or departing the airport.

(3) Speak slowly and distinctly. If you do not get an immediate reply, wait a few moments and repeat your request. Please note that pilots announcing their departure are not normally acknowledged.

(4) State the facility or airport you are calling, your ultralight identification, your location relative to the airport, and your intended operation. Example: "Leesburg Ultralight 12593U is 5 Miles North, Landing."

(5) If you still do not get a reply, proceed cautiously toward the airport. If departing the airport, be careful to visually clear the area in *all* directions prior to entering the takeoff area. Remain on the proper radio frequency and listen for any aircraft which may be in the area.

(6) Once you have completed your landing or have exited the area, it is good practice to let other aviators know that you are no longer airborne in the vicinity of the airport. Example: "Leesburg, Ultralight 12593U is Clear of the Runway" or "Leesburg, Ultralight 12593U is 2 Miles South, Leaving the Area."

## WEATHER INFORMATION

The desire to leave the ground and explore the world from the air has inevitably tied you to weather and its effect upon you. No pilot, amateur or professional, can safely attempt a flight without considering the present and expected weather conditions. Weather is a factor in most aviation accidents. It cannot be emphasized too strongly that if you are to continue to operate safely, it is essential to know and understand the environment in which you are flying.

Individual pilot weather briefings from FAA flight service stations are provided to pilots on a "first come, first served" basis. The number of briefers available today is insufficient to meet user demands without the prospect of considerable delays. The FAA is taking steps to remedy this. An automated system currently under development is designed to accommodate direct user access and will be able to provide increased services. Until that system is operational, the present FAA flight service system may not be able to accommodate all the needs of ultralight fliers.

**SOURCES OF WEATHER INFORMATION.** Many sources of weather data are available to aviators. The following sources will assist you in acquiring and evaluating as much weather data as possible.

National weather is broadcast weekdays in a live 15 minute television program called AM Weather. The program is carried by about 250 public broadcast stations in the early morning. This program features meteorologists from the National Weather Service and the National Environmental Satellite, Data, and Information Service (NESDIS). They use the latest guidance and data available to produce a thorough program. The program's surface and forecast maps, satellite imagery, radar maps, and upper air charts, along with the hazardous weather watches, are ideal for acquiring broad scale weather information. Consult your local television schedules to obtain time of broadcast in your area.

Many cable TV systems now include 24 hour weather channels. Some of the programs include aviation weather.

Transcribed Weather Broadcasts (TWEB) for aviation are made on numerous FAA VHF omni-directional rages (VOR), nondirectional radio beacons (NDB's), and at selected airports that provide automatic terminal information services (ATIS). These transcribed broadcasts are continuously updated during their hours of operation.

Broadcasts over radio beacons are made in the range of 200-400 KHz and can be received on relatively inexpensive radio receivers. VOR and ATIS broadcasts are made on VHR aviation radio frequencies between 108-136 MHz. There are many moderately priced radios available that will receive these frequencies.

The content of TWEB and ATIS broadcast in some cities can be received over the telephone. The telephone numbers to use can be found in the telephone directory under United States Government, Department of Transportation, Federal Aviation Administration. TWEB recordings will be listed under Air Traffic Control Tower.

On nondirectional radio beacons and selected VHF omni-directional ranges (VOR's), the broadcasts may include synopsis, adverse conditions, route forecasts, outlook, winds aloft forecasts, radar reports, surface weather report, etc.

Broadcasts on other VOR's may include only surface weather reports, terminal (airport) forecast for the local airport, adverse conditions, etc.

ATIS broadcasts may include local ceiling, visibility, obstructions to vision, temperature, wind direction (magnetic) and speed, altimeter setting, etc. The information is applicable only to the airport located at the ATIS site, but it may be used in evaluating the trend of existing weather.

All the above facilities and their frequencies may be identified by studying sectional aeronautical charts that are sold at many airports. Much of the same information is found in the U.S. Government Flight Information Publication, Airport/Facility Directory. Comprehensive explanations of all these services are printed in the FAA Airman's Information Manual (ATM). The publication is available through the U.S. Government Printing Office. Other excellent sources to find out frequencies and what is available, are pilots handbooks published by user organizations.

In most large metropolitan areas, the National Weather Service provides continuous broadcasts of local weather conditions on two frequencies that can be received by inexpensive radios available at many retail outlets.

Pilots Automatic Telephone Weather Answering Service (PATWAS) is available in most large metropolitan locations. This is a telephone recording of local and route weather information that can be obtained by dialing a telephone number found under the same heading in the phone book as listed above for TWEB.

If you live in the Washington, D.C. or Columbus, Ohio areas, you should become familiar with the voice response system (VRS) installed at these locations. This is a computer based test system that provides weather data over the telephone. The user needs only to have a "TOUCHTONE" phone to access the system. Since this is a test system, the products available may vary. The latest information available and directions on using this system can be obtained by sending a stamped self-addressed envelope to:

Voice Response System
DOT/FAA/FAATC
ACT 110
Atlantic City, N.J. 08405

These many sources of weather data are only part of a safe weather operation. Other factors include a knowledge of how to interpret the weather data correctly, and when to exercise good judgment and not fly. There are many Government and civil sources that supply educational material on weather and user organizations are developing courses aimed at improving the ultralight operator's understanding of weather. One of the best efforts ultralight operators can make in their own behalf is to find out about weather. Many members of the aviation community have learned that weather, above all other aspects of our environment, is irreverent of even the most experienced aviator.

**MICRO-METEOROLOGY.** While the list of available weather information is impressive, it may not provide the ultralight operator with the actual weather and wind conditions at the operating site. One of the most critical factors in conducting a safe takeoff and landing is accurate information of the wind conditions on the surface. There may be many indications of what the wind conditions are at the flying site. The information provided herein is designed to assist you in understanding and using those indicators.

**Wind Direction.** One of the best indicators of wind direction near the surface is derived by the use of a windsock or wind streamers. The direction of the wind is clearly indicated, as is the velocity. Because ultralight vehicles are very susceptible to wind, we recommend that several windsocks or streamers be located around the landing site. Another means of learning the wind direction on the surface is from nearby ponds or lakes. The "glassy" or smooth water area along the shore indicates the direction from which the wind is blowing. The further out into the body of water the glassy area protrudes, the lower the wind velocity. Be careful when using this method that the shoreline is not subject to major obstructions such as high trees or a steep, high bank. Yet another indicator of wind direction and velocity is the natural vegetation such as tall grass, trees, and bushes. Caution should be used here too, for the trees themselves can cause the wind direction to change significantly, see **Turbulence and Wind Shear** below. Other indicators of surface wind are smoke and blowing dust. Learn to use them all and learn to cross check the information of one against the other. They are inexpensive resources that may save your life.

**Wind Gradient and Gusts.** Wind gradient is change in the velocity of the wind with an increase/decrease in altitude. Normally, wind velocities will increase as the altitude increases. Conversely, because of the drag effects of the earth, winds may significantly decrease as you get closer to the ground. If the winds decrease at a faster rate than can be accounted for by pitch and thrust changes, the vehicle may enter a stall. For this reason, when descending or climbing in close proximity to the ground, a safe margin of extra airspeed is recommended. Also affecting the ultralight vehicle are wind gusts. The danger inherent in gusting wind conditions is amplified during the takeoff and landing phases of flight. A sudden gust of wind could lift the ultralight up quickly, only to abandon the pilot 20 feet above the ground. The result is often a stall. Another effect of gusting winds is the effect on the airframe of the vehicle. Strong gusts could easily and quickly exceed the design limits of the vehicle, especially if the pilot is performing a maneuver which is already putting some "load" on the airframe. The best advice for operating in gusting winds is to ask yourself,

"Do I really need to be doing this?" If you absolutely, positively have to be there, fly gently and maintain extra airspeed during the takeoff and landing. Fly the vehicle right down to the ground with a minimum landing flare, and, after you've landed, ask yourself, "Do I really want to do that again?"

**Turbulence and Wind Shear.** The most critical altitudes for micro-wind changes are between 30 and 75 feet above ground level. This depends, in part, on the nearness of the surrounding obstructions such as large trees, buildings, and hills. The effect of these obstructions is often turbulence or a sudden change in wind direction and velocity often referred to as wind shear. Turbulence can be especially dangerous in ultralights due to their light weight. Ground turbulence consists of vortices and eddies, vertical blasts of air, and rotors (dust devils). Turbulence is caused by winds moving across and around objects, and by thermal heating of the earth's surface. Wind shear can result in a sudden reduction in the relative wind over the vehicle's lifting surfaces. When this happens, the vehicle may very quickly enter a stall. At low altitude it may be nearly impossible to recover in the distance remaining to the ground. Because of the effects turbulence and wind shear have on the safety of ultralight operations, it may be wise not to fly ultralights in winds exceeding 15 mph. And even then, there will be some circumstances when 15 mph is too much. Also, keep in mind not only your own piloting skills, but the abilities of your craft to handle a crosswind during takeoff and landing. If you are in doubt, err on the side of safety and leave the enjoyment of flying for another time, perhaps another day.

## ACCIDENT INFORMATION AND OTHER SOURCES

### NATIONAL TRANSPORTATION SAFETY BOARD (NTSB).

The NTSB is the official Government investigator for all transportation safety issues. Its purpose is to impartially analyze occurrences which may indicate a transportation safety problem and to recommend corrective action. The NTSB has decided to investigate all fatal powered ultralight vehicle accidents and other selected ultralight accidents and incidents which may involve significant safety issues. The Safety Board will also investigate ultralight vehicle accidents impinging on civil aircraft operations or on persons and property on the ground. The Safety Board will review accident data and the safety efforts of the aviation community in order to keep abreast of any emerging safety problems and will be available to provide technical assistance in remedying those problems.

IF YOU ARE WITNESS TO OR INVOLVED IN AN INCIDENT/ACCIDENT INVOLVING THE OPERATION OF AN ULTRALIGHT VEHICLE, NTSB REQUESTS YOU DO THE FOLLOWING:

(1) Immediately attend to the medical and physical needs of the situation. Notify the local authorities if assistance is needed.

(2) Do not move or remove any debris associated with the occurrence.

(3) Write down as much as you can remember. This will be very helpful in accurately recalling the incident.

(4) Notify, or have the local authorities notify, the nearest NTSB Field Office. This information can be found in the local phone book under U.S. Government, National Transportation Safety Board, or call your local FAA office and request the NTSB telephone number.

(5) If you are able, take photographs of the site, and get the names and phone numbers of any witness.

NTSB requests that you be very helpful in reporting such incidents as this will give all of the owners/operators of ultralights a chance to benefit from the knowledge gained during the investigation. The Safety Board investigation is fact-finding in nature and will not be used to substantiate any violation of Federal Aviation Regulations.

Additionally, the FAA supports the goals of private organizations and associations to provide technical and operational assistance to the ultralight industry in enhancing the reliability of the vehicles and the safety of the sport. The FAA encourages all participants in the sport of ultralight flying to report any incident, accident, structural or mechanical failure of an ultralight to the private organizations and associations actively representing the sport.

### AIRPORTS DISTRICT OFFICE (ADO).

Airport district offices are located throughout the country and serve a specific geographical area. Their primary purpose is to assist the aviation community and state and local governments in the planning and development of landing facilities. Under FAR 103, ADO's would be your best source for information pertaining to the establishment of a flightpark and the environmental considerations associated with operations.

For the phone number and location of the ADO serving your area, consult your local phone directory under Department of Transportation, Federal Aviation Administration, Airport District Office or Regional Airport District Office.

### FAA AIR TRAFFIC CONTROL FACILITIES.

There are three major types of air traffic control facilities with which you may come in contact. The following information should assist you in determining which one to call.

**Flight Service Station (FSS).** The Flight Service Station's primary function is to provide the pilot with preflight weather briefings and also Notices to Airmen (NOTAMS) which have information as to the status of airports and facilities; the conduct of special activities (parachuting airshows, military exercises, etc.); and the presence of known temporary structures such as a crane located near an airport. For the ultralight operator, FSS's can be a means of obtaining guidance on which FAA facility could best be of assistance. For the role FSS's play in providing weather information to ultralight pilots, see Chapter 3.

**Air Traffic Control Tower (ATCT).** There are many air traffic control towers located throughout the country. Each serves a particular airport and provides pilots with information on the movement of other aircraft in and around the airport. In some circumstances, ATCT's have an approach control associated with them which provides separation between aircraft over a wider geographic area. Under FAR 103, ATCT's would be your contact point for operations in an airport traffic area. In many instances, operations at nearby airports with control zones may also be coordinated through the nearest ATCT.

**Air Route Traffic Control Center (ARTCC).** There are 20 ARTCC's located around the country. Each one covers a very large geographic area and provides radar separation services to aircraft through the use of remote radar and radio communication facilities. In some areas, the ARTCC functions are an approach control and has responsibility, under FAR 103, for providing authorization for ultralight operations in a control zone. Due to the size and vast area of coverage of ARTCC's, it is better to contact the FSS or ATCT nearest you for assistance in obtaining required authorizations.

For the phone numbers and locations of the FSS, ATCT, or ARTCC you wish to call, consult your local telephone directory under Department of Transportation, Federal Aviation Administration. Each facility should be listed separately: Flight Service Station; (airport name) Air Traffic Control Tower; and Air Route Traffic Control Center.

**GENERAL AVIATION DISTRICT OFFICE (GADO).** These offices are located throughout the country and are staffed by Flight Standards personnel. Their primary purpose is to serve the general public and aviation industry on all matters relating to the certification and operation of general aviation aircraft. These responsibilities include accident prevention programs, general surveillance of operational safety, and the enforcement of FAR. Under FAR 103, GADO's are your best source of information for items such as vehicle applicability, hazardous operations, and operations over congested areas. Should you desire, GADO's can also provide you guidance and assistance in certificating your ultralight as an aircraft.

For the location and phone number of your nearest GADO, consult your local telephone directory under Department of Transportation, Federal Aviation Administration, General Aviation District Office or Flight Standards District Office.

**PUBLICATIONS.** The Federal Government and the aviation industry have devoted considerable energies to producing informational and training publications which are invaluable to pilots. Listed below are some of the publications available from the Superintendent of Documents, U.S. Government Printing Office, Washington, D.C. 20402.

Other sources of useful information can be obtained through the various organizations, manufacturers, and associations working within the aviation community.

**Airmen's Information Manual (AIM).** This manual contains the basic fundamentals required for safe flight in the U.S. National Airspace System. It includes chapters on navigation aids, airspace, air traffic control, flight safety, and good operating practices. It also includes a pilot-controller glossary. The AIM is issued every 112 days and the annual subscription price is $17.

**Pilot's Handbook of Aeronautical Knowledge.** This handbook contains essential information used in training and guiding pilots. Subjects include the principals of flight, airplane performance, flight instruments, basic weather, navigation and charts, and excerpts from flight information publications. This handbook is one of the most complete sources of aeronautical information available. The current price is $10.

Listed below are some of the publications available from the FAA.

**Flight Standards Safety Pamphlets.** These pamphlets are used in the General Aviation Accident Prevention Program and are produced primarily to be distributed at accident prevention seminars by GADO personnel. Titles available include: Density Altitude, Weight and Balance, Propeller Operation and Care, and Planning Your Takeoff. There are many other subjects available. Pamphlets may be obtained in reasonable number at no charge from the FAA Accident Prevention Specialist assigned to your local GADO.

**FAA Advisory Circulars.** The FAA issues advisory circulars to assist and inform the public on matters affecting aviation. Advisory circulars are issued in a numbered-subject system corresponding to the subject areas of the FAR.

For example, this advisory is numbered AC 103-6 because it deals with information pertaining to FAR 103 operations. There are more than 400 free advisory circulars available. Subjects which may be of interest to the ultralight operator include:

| | |
|---|---|
| AC 60-4A | Pilot's Spatial Disorientation |
| AC 90-23D | Wake Turbulence |
| AC 90-42C | Traffic Advisory Practices at Uncontrolled Airports |
| AC 90-48B | Pilot's Role in Collision Avoidance |
| AC 91-36B | VFR Flight Near Noise Sensitive Areas |

For a complete listing of all available advisory circulars, send your request for the Advisory Circular Checklist, AC 00-2 to:

U.S. Department of Transportation
Subsequent Distribution Unit,
  M-442.32
Washington, D.C. 20590

Please enclose a self-addressed mailing label to expedite the processing of your request.

Additionally, the FAA publishes numerous other documents dealing with a variety of subjects. The Guide to Federal Aviation Administration Publications lists the information available from the FAA and also provides a list of civil aviation related publications issued by other Federal agencies. A free copy of this guide is available from the address listed above.

**Airport/Facility Directory (A/FD).** Issued every 8 weeks, the Airport/Facility Directory is a civil flight information publication which contains a directory of all airports, seaplane bases, and heliports open to the public. Available from the National Ocean Service, NOAA Distribution Branch, N/CG33, Riverdale, Maryland 20737, the directory includes information on communication frequencies, navigational facilities, and certain special notices such as curfews. Directories are sold on a single copy or subscription basis and cover a specific geographic area of the United States, Puerto Rico, and the Virgin Islands.

# MY FIRST FLIGHT IN AN ULTRALIGHT

**OSHKOSH FLY-IN** hosted by the International Experimental Aircraft Association. ULTRALIGHTS fly past the main entrance. This was the most exciting aviation event I've ever witnessed.

**BEEACHCRAFT. STAGGERWING** taxis to the flight line.

**P-38 WARBIRD** lifts off for a fly-by demonstration of W.W.II fighter tactics.

**CANADIAN SNOWBIRDS** exhibit close formation precision flying.

**QUICKSILVER MX** equipped with floats makes a landing approach at Lake Winnebago, Wisconsin.

**TINA TREFETHEN**, Pilot-in-Command with the Eipper Flight Team invites me for a flight in the QUICKSILVER MX. *Below Left:* Tina on the flight line in her flight suit and parachute harness. *Below Right:* Airborne, I photograph Tina and the carabiner that secures her parachute to the ULTRALIGHT's Root Tube to safely bring plane and passengers to earth in an emergency.

| | |
|---|---|
| Date: | Monday 22 August 1983 0700 hours |
| Weather: | Slight overcast |
| Temperature: | 63° |
| Wind: | Westerly, 3 to 5 knots |
| Place: | American Aerolights Dealer Training Center Coronado Airport Albuquerque, New Mexico |
| FAA-Certified Flight Instructor: | Robert Mullikin |
| Chief Flight Instructor: | Kris Williams |
| Aircraft: | Eagle XL |
| Training: | FLIGHT SIMULATOR – TAXI-TOW TRAINING |
| Student: | Rick Carrier |

**CORONADO AIRPORT.** Looking East, 0700 Hours.

# YOUR FIRST STEP TO FLIGHT IN AN ULTRALIGHT

## DAY 1

"That's Coronado Airport over to our right, Rick. This is where you'll take your flight training."

Robert Mullikin's voice was soft and calm, quite unlike the edgy excitement surging through me in anticipation of the beginning of a lifelong desire to fly. We were cruising north along Interstate 25, that led to Colorado if we continued. My eyes were fixed on the top of the Sandia Mountains, miles off to the east. The sun was coloring the base of the deep grey clouds floating above the crest in a warm orange glow. The ground along the base was obscured by wispy fog that hovered above it all the way to the airport. I kept my eyes on the mountains as I spoke.

"This airfield is much larger than I expected."

Bob smiled, glanced at me, and in a mellow voice that crackled with humor, asked,

"What were you expecting? A grass strip cut in the middle of the desert?"

"Sort of," I answered. "I wasn't expecting to see so many airplanes at an Ultralight airfield. From all the stories I'd heard at Oshkosh, I was under the impression the people in general aviation were against ultralights."

Bob spoke as he smartly wheeled the car onto the airfield access road. His voice was serious.

"The general impression the public has is that general aviation is against ultralights. Don't believe it. A few complain, mostly because ultralight pilots don't have to go through all that training and expense licensed pilots must do to become licensed. It sort of galls them that ultralight flying is so easy, and you can fly without a license. But the Cessna dealer here is a friend of ours. He's a good guy, as are most of the serious pilots here and around the country. General aviation has really taken it on the nose financially because of the recession. The ultralight movement is bringing new life into aviation, so don't

33

**XL EAGLE** out on an early morning flight at Coronado Airport near Albuquerque, New Mexico.

listen to the complainers. They're always around; no matter what they say, ultralights are here to stay."

"That was my impression also. But the media picks up on all the negative aspects of the ultralight adventure, and that is what the public reads and remembers. That's one of the reasons I'm doing this book: to let the general public see and experience flight in an ultralight. Also, Bob, I spent a long time studying, reading and inquiring about the manufacturers' reputations in aircraft assembly and safety standards, all the way to their training and dealer requirements. That is why I am here, to be taught flying by the best in the business. All the pilots and manufacturers I spoke to, at Oshkosh and on the phone, when they understood that I was going to learn flying from scratch, with no previous experience in powered aircraft, advised me to start with the Eagle XL, then move on to the Quicksilver MX, in that order, before attempting flight in the higher performance aircraft. Even your competitors, when pressed, suggested this approach. Your company has the dealer network, care and detail in safety standards, manufacturing and training skills required all across the country. This impressed me; that's why I'm here."

Bob smiled again. His voice was back to mellow. "It's nice to know that other people think we are as good as we work hard at to be. I say this because everyone here at American Aerolights demands the highest standards from every aspect of our training, manufacturing, facilities, dealer selection and maintenance, and, most importantly, the people we hire to represent us. We have a selection program that is the best. Everyone here at our company that deals with the public are experienced pilots. Larry Newman, our president, has very high standards, we respect and follow his lead all the way."

"I'm looking forward to meeting Larry and the rest of your company."

I looked up at the rotating beacon high on its metal tower sweeping around and around, sending brilliant beams of light into the sunrise. As my eyes pulled away from the beacon, they fixed on two specks in the sky far away towards the Sandia Mountains. "Those two spots in the sky out there look like ultralights." Bob hunched forward and glanced over to the mountain.

"Probably Romuald Drlik and somebody else. Romuald, by the way, is the original designer of the Eagle. Probably on a test flight. He flies every chance he gets. You'll meet him later. He works on new designs here. Early mornings are the best time to fly ultralights. Very light winds. That's why we schedule our training sessions this early. By mid-afternoon the ground has heated up and the rising thermals make flying choppy and choppy air is not good for training, or—" Bob's smile returned, "for fledgling flyers like yourself, Rick."

His words made me feel welcome, and dissipated the feelings of anticipation that had jangled my nerves earlier.

As Bob wheeled the car around the end of a low white building, the American Aerolights Training Center, my curiosity flamed. Kris Williams, whom Bob had said earlier would be my flight instructor, was perched on a low platform that was attached to the front end of a station wagon. He was busily attaching a maze of steel cables to the nose and wings of an Eagle XL. The image resembled a massive spider web that had snared a big colorful dragonfly. I turned to Bob. My voice rose up in pitch. "Is that the flight simulator?" Bob grinned as he hopped out of the car. "That's it, Rick; come on, I'll introduce you to Kris."

I liked Kris Williams immediately. He was young, serious, very friendly, and had a smile that warmed the heart. After introductions, Kris pointed to the simulator.

"While you get a cup of coffee, I'll finish setting up the simulator. Then we'll go out on the runway and give you a chance to fly the Eagle."

Reluctantly, I followed Bob into the offices.

I glanced at my watch. It was seven-ten. The offices were luxurious. Having visited the American Aerolights tent at Oshkosh a few weeks earlier, where I had first met Bob to arrange for this training, had not quite prepared me for their training center. I was impressed. It was very clear to me that these people were first rate, the top of the line. I felt secure here, and knew that I would receive the very best instruction possible. Bob had said earlier that all members of the team here were certified pilots, and that each and every one of them had been carefully screened by the management to ensure competency in all matters concerning flight. This fact was becoming more obvious every moment.

Tastefully arranged on the walls and tables of the reception room were signed photographs and gleaming awards that had been presented to two of the American Aerolight top management, Larry Newman, the thirty-six-year-old company president, and Bryan Allen, thirty-one, vice president of public affairs.

While Bob went to fetch a cup of coffee for me, I nosed up to the photos. There was Larry Newman, high in the sky, on his historic transatlantic crossing in Double Eagle II, a balloon piloted by two of his Albuquerque business associates, Ben Abruzzo and the late Maxie Anderson, along with photos of his last transpacific crossing in Double Eagle V in 1981 with Rocky Aoki of the Benihana restaurant chain. Clustered around Larry's pictures were several photos and pictures of Bryan Allen's 1979 world achievement of flying the Gossamer Albatross on the first man-powered flight across the twenty-six miles of the English Channel. Bryan won the largest aviation prize ever given to a person for that flight.

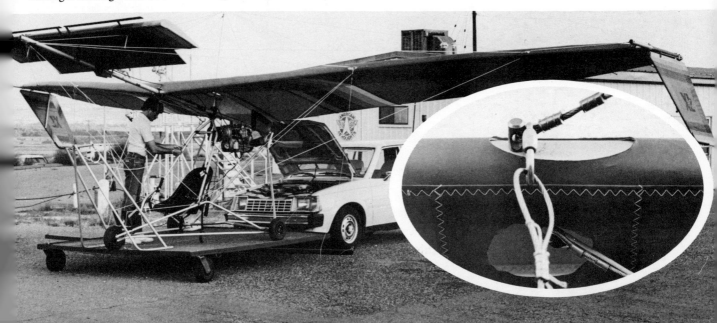

**KRIS WILLIAMS, CHIEF FLIGHT** Instructor at the American Aerolight Training Center at Coronado Airport is shown here mounting the EAGLE XL to the flight simulator. The inset shows how the simulator's restraining cables are attached to the aircraft's wing wires.

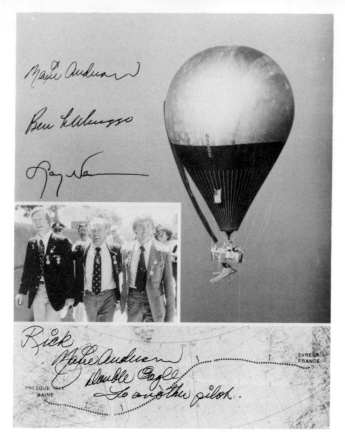

**DOUBLE EAGLE II AND HER CREW.** The late Maxie Anderson is on the left. In the center is Ben Abruzzo and on the right Larry Newman, President of American Aerolights, Inc. Here they are shown celebrating their historic transatlantic crossing. The map below is the route they flew the balloon across the Atlantic Ocean, and their landing site in France. Larry Newman's hang glider, dangling below the gondola, had to be jettisoned into the sea to lighten the balloon's ballast.

**BRYAN ALLEN, VICE PRESIDENT** of Public Affairs for American Aerolights, Inc. is pictured below, sealed in a plastic covered cockpit of the Gossamer Albatross, the first man-powered aircraft to fly across the English Channel. This world-record flight

Bob came back into the carpeted reception room and asked if I was ready. Quickly swallowing the hot coffee, I rose up from a plush leather sofa, put the book *Double Eagle II* back on the polished wooden coffee table, loaded my camera and recording gear, and said, "Let's go. I've been waiting for this moment for a long time."

When we stepped outside, Bob looked up at the sky. The morning ground fog had cleared. The sun was higher over the mountain crest. The air was calm.

"High cirrus clouds. This will be a nice calm day for your first lesson, Rick."

took two hours and forty-nine minutes of excruciatingly painful pedaling. His average speed over the water was eight MPH (miles per hour). The prize for this epic flight was $200,000. The plane is now with the Smithsonian Air and Space Museum.

**ROBERT MULLIKIN (BOB)** is an FAA Certified Flight Instructor/Airplane Instruments, and a sales representative for American Aerolights. He conducted the classroom instructions in Aerodynamics and Meteorology along with guidance throughout all flying lessons.

**KRIS WILLIAMS, CHIEF FLIGHT INSTRUCTOR**, his wife Rhonda and his daughter Kristin. Kris is in charge of all of the flying lessons at the American Aerolights Training Center.

ABOVE IS THE SAIL LOFT of American Aerolights. It is as clean as an operating room and for good reasons. Any dirt or foreign matter left in the fabric seams, surfaces or structural parts will cause abrasions and wear on those parts. Here in this room everyone wears slippers and like all the rooms and offices, positively no smoking is allowed.
**BELOW IS THE WING FRAMING ROOM** where the wings are aligned and trimmed before the aircraft are shipped to their dealers.

# AIRCRAFT PREFLIGHT TECHNIQUE

"We're going to start with our preflight, which is the first thing you do every time you get ready to fly.

"We always start at one position to develop good habits. You never start here, then there, and at another place later. By always starting at the same place you never forget where you ended.

"We look at and physically feel all the nuts, bolts, and safety washers. All the safety pins you look at, touch and study. It is the development of good habits to touch every safety pin. That way you get both a visual and physical inspection.

"Check the articulating canard. Check for washer placement — clevis pins — here, on the tip rudder, we have four washers.

"Walk out and around the aircraft counterclockwise — clockwise from the pilot's point of view. Check the bell crank and teleflex cables stops, making certain the bridges are locked.

"Make certain all safety pins are connected. Safety pins have a tendency to disengage during towing.

"Work the foot pedal operation of the tip rudders — full deflection and also the nosewheel steering function.

"Operate the control surface system. In the canard movement, a full up and down.

"Next, this pop rivet, and then move down the teleflex cable. Notice everything that touches the cable for abrasions or wear.

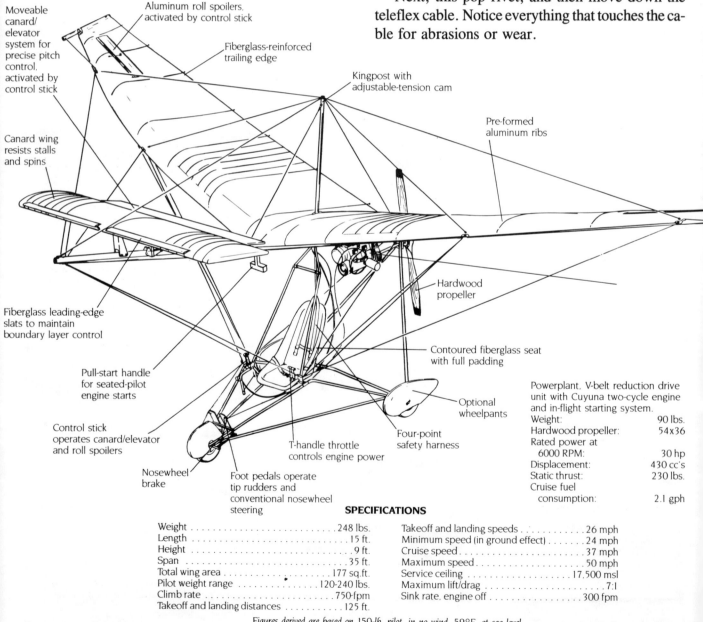

**SPECIFICATIONS**

Weight . . . . . . . . . . . . . . . . . . . . . . . . . . . 248 lbs.
Length . . . . . . . . . . . . . . . . . . . . . . . . . . . 15 ft.
Height . . . . . . . . . . . . . . . . . . . . . . . . . . . 9 ft.
Span . . . . . . . . . . . . . . . . . . . . . . . . . . . . 35 ft.
Total wing area . . . . . . . . . . . . . . . . . . . 177 sq.ft.
Pilot weight range . . . . . . . . . . . . . . . . . 120-240 lbs.
Climb rate . . . . . . . . . . . . . . . . . . . . . . . 750 fpm
Takeoff and landing distances . . . . . . . . . . 125 ft.
Takeoff and landing speeds . . . . . . . . . . . . 26 mph
Minimum speed (in ground effect) . . . . . . 24 mph
Cruise speed . . . . . . . . . . . . . . . . . . . . . . 37 mph
Maximum speed . . . . . . . . . . . . . . . . . . . 50 mph
Service ceiling . . . . . . . . . . . . . . . . . . . . . 17,500 msl
Maximum lift/drag . . . . . . . . . . . . . . . . . . 7:1
Sink rate, engine off . . . . . . . . . . . . . . . . . 300 fpm

*Figures derived are based on 150-lb. pilot, in no wind, 59°F, at sea level*

## PREFLIGHT

Start at nose and walk around Eagle, returning to nose.
1. Check all bolts installed and safetied in canard, slats tight, canard mechanism operating smoothly.
2. Check nosewheel area; all cables and nuts safetied.
3. Check stick and seat; teleflex cables mounted and safetied, seat adjusted for leg length, support tube safetied.
4. Check spoiler bellcrank area.
5. Walk along leading edge; no twisted thimbles, rudder and spoiler cables, routed through pulleys and free.
6. Check tip rudder, control cable safetied, rudder secure, check for free movement and functioning of both stops.
7. Inspect roll spoilers; velcroed, cable safetied, both spoilers 2 inches up.
8. Check trailing edge; all balls in place, trailing edge batten installed.
9. Check power system; bolts tight & safetied, belt tension, prop unblemished, bearings free, fuel tank fitting in place, fuel level OK.
10. Repeat steps 5, 6, 7 & 8 for opposite wing.
11. Check landing gear; triangle bar clevises installed, tires OK, bolts safetied, throttle cable and handle OK.

**PRE-TAKEOFF CHECKLIST**
1. Wear a helmet, chinstrap secure.
2. Wear or check parachute, if ballistic type.
3. Clear prop, prime and start engine.
4. Fasten belts, let engine warm 1-2 minutes.
5. Taxi to run-up area; check stick pitch movement, rudders free, roll spoilers free.
6. Check safety belts fastened and tight, apply brake and run up engine for ground check.

**TAKEOFF**
1. Apply takeoff power.
2. Reduce power several hundred rpm for sustained climbs over one minute.
3. Fly safely and ENJOY.

## ULTRASPORT SRL PURCHASES EAGLE RIGHTS

Wings hand-crafted with four-ounce Dacron

Tip rudder, activated by foot pedal, rainbow-spectrum Dacron

On February 2, 1984, Romuald Drlik, designer of the Eagle ultralight, sold the manufacturing rights to produce the Eagle, Eagle XL, and any possible variations to Ultrasport SRL, an Italian-based ultralight manufacturer. Rights to construct the Eagle previously were held by American Aerolights, Inc. Mr. Drlik revoked these rights prior to selling them to Ultrasport SRL. This action regarding the manufacturing rights for the Eagle and its derivatives was part of an agreement between the management of American Aerolights, Inc., Romuald Drlik, and the President of Ultrasport SRL, Luigi Accusani.

It is Ultrasport SRL's intention to manufacture the Eagle ultralight and its variants for distribution throughout Europe and the United States. With the current value of the Lira versus the Dollar, Accusani stated he is confident the Eagle can be manufactured in Italy, shipped to the US, and sold for less than if it were still manufactured in the USA.

Along with manufacturing rights, Drlik provided Ultrasport SRL all specifications, drawings, test data, patterns, and other materials to allow Ultrasport SRL to begin complete and total manufacture of the Eagle, Eagle XL, and Eagle 2-Place. Although no tangible value can truly be placed on this information, it would be nearly impossible for any company to correctly produce the Eagle line without it. Technical design and support will be provided by Drlik on a consultant basis to Ultrasport SRL. Royalties for each Eagle produced will be paid to Drlik, the Eagle's designer, by Ultrasport SRL just as they were being paid formerly by American Aerolights.

Although Ultrasport SRL will soon be producing spare parts for the Eagle line, there is an additional parts supplier based in North America who will also be providing parts service for the dealerships and owners of ultralight aircraft built by American Aerolights, Inc. More information will be coming soon on this parts supplier.

ULTRASPORT SRL, GIESSE TEAM, FRAZIONE-MONBERNO 10060.
GRAZIGLIANA, TORINO, ITALY 011-391215-413-845

---

# WARNING

**FLIGHT LIMITATIONS OF THE EAGLE ULTRALIGHT.** Factory authorized pilot training is required prior to powered solo flight of this aircraft. This Eagle ultralight is designed for maximum keel tube angles of + 42° or − 18° to the horizon, maximum bank angles of 60° to the horizon and maximum dive speeds of 55 mph. Minimum flight speed of the Eagle ultralight is 20 mph. **THIS AIRCRAFT IS NOT APPROVED FOR INVERTED FLIGHT NOR AEROBATIC MANEUVERS.** Flight outside this maximum and minimum envelope is hazardous and should not be attempted. Flight in the Eagle ultralight involves travel in three dimensions, and such activity is subject to mishap and injury including death. Pilot acknowledges that there are no express warranties, no implied warranty of merchantability, and no warranty of fitness of any Eagle ultralight for any particular purpose, and that the flight is attempted solely at pilot's own risk.

# PREFLIGHT

"Move the spoilerons up and down. They should work smoothly.

"On the landing gear, you look for things that *don't* look like they ought to be.

"Then work your way down the leading edge of the wing, running your hand over the fabric, feeling for dents, and observe the routing of the cables and pulleys to the rudder, while touching all the safety pins and looking at all the pieces. Everything is visual; nothing is hidden.

"Wing battens, if extended, push them back into place, and seat the little balls in the ends of the battens.

"King post—all fittings and connections secure.

"Wire tension—if loose, tighten to manufacturer's specifications.

"Propeller should always be horizontal.

"Check the aircraft for symmetry. If something is not symmetrical, it is an indication that something has been changed and is not quite the way it should be. If there are three washers on one side, make sure there are three on the other side.

"When in doubt, do not fly the aircraft until your questions have been answered by someone who knows the problem, and the problem is corrected. Remember, Rick, when you fly an airplane, you are totally responsible for its airworthiness. If a cable pops off a connector without a safety pin, and you overlooked it because someone forgot to put one on, it's nobody's fault except yours, not the individual that neglected the attachment. Your safety depends on a careful, knowledgeable preflight every time before you sit in the pilot's seat."

When I was examining the teleflex cable, I noticed it was missing a lock ring. I turned to Kris and pointed this out to him. He pulled his hand from his pocket and handed me the missing ring. I replaced it and continued my preflight. When I had finished, being back at the nose wheel where I had begun, I turned to him asking,

"I've completed my preflight. Did I miss anything?"

"No, that ring was the only piece I removed."

He then went to an envelope attached to the Eagle XL bearing the notice: "This aircraft cannot be flown." Inside the envelope was a description of what needed to be done to make it flyable."

Kris removed it and told me a story.

"A student of mine was preflighting an Eagle and when he was certain the plane was ready for flight, and that he had checked everything, I asked him, 'Are you positive you checked everything on the preflight check list?' He quickly said that he had. So I pulled my hand from my pocket, fist closed, took his hand and dropped five missing safety clip pins and rings into it. The student's face went slack as I said to him, 'Before you fly that plane, you'd better put these back in their proper places. When you finish finding where they go you will have preflighted the aircraft properly.' A couple of hours later we continued on with his training."

A focus in my mind sharpened on that story and planted it deep into my recall memory so it would be a constant reminder, never to get caught without making a thorough preflight inspection, with emphasis on close, painstaking attention to details, without distractions of any description to my preflight procedures. This was my first of several mental disciplinary lessons I would learn that touched home.

RESPONSIBILITY

To yourself and all the others about you as you fly into the sky. Everything you do when you are around the aircraft is vital to the safety of not only the aircraft but also to your very life. So therefore RESPONSIBILITY is the first discipline lesson, which leads directly to the next. You are completely RESPONSIBLE for the truth in reality of the life and death results of your preflight OBSERVATIONS and FEELINGS, by your hands, intuitions, faith or whatever you choose to aid yourself in making a personal commitment to the safety of the aircraft before you use it.

---

**Remember this and burn it into your mind:**

I. ALWAYS use the PREFLIGHT CARD everytime you preflight your aircraft.

II. Accept total RESPONSIBILITY for your preflight.

III. OBSERVE everything on or about your airplane intensely.

IV. FEEL every part as you look at your aircraft.

If you don't follow this advice and you miss something on your preflight, you've placed your bid to buy the farm.

**A. THE NOSE PLATE** on the tip of the bowsprit is a critical junction of the EAGLE XL's cables. This Nose Plate secures the King Post wires, Canard Stabilizing Cables and the Lower Flying Wires. If the Safety Pin, holding the wing nut in place is missing, the wingnut could vibrate loose while in flight and cause a disaster.

**B. THE ARTICULATING CANARD SAFETY PINS** and the Teleflex *safety pin* are clearly visible and should be checked carefully. Any malfunction here can cause a crash.

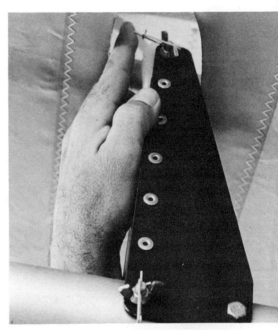

**C. THIS SAFETY PIN** holding the wing nut in place on the Canard pilon keeps the Canard Wing in place. If the wing nut came off in flight the aircraft will pitch forward and descend vertically.

**D. THE TELEFLEX ROD END BEARING** that operates the Spoilerons is improperly amounted. The Rod End Bearing should be above the Bell Crank Lever and the Safety Pin below. **E. HERE IS ONE OF THE REASONS.** When the Safety pin is removed the Rod End Bearing can slip right off. Although a malfunction of the Spoilerons during flight will not necessarily cause a crash, any malfunction should be ruthlessly avoided by a thorough Preflight Inspection.

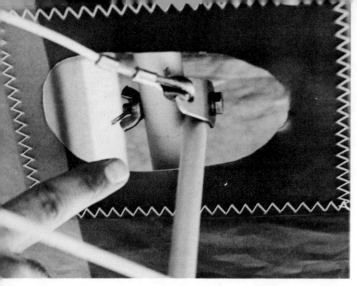
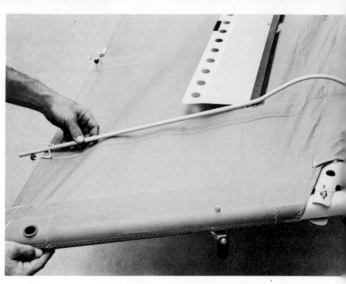

**F. THE JUNCTION WHERE THE TANG** is bolted to the Compression Strut and Verticle Strut is secured by a Wing Nut and held in place with a Safety Pin. It is vital Preflight inspection area.

**G. THIS ALUMINUM WING RIB** fits into the Wing Rib Pocket located directly underneath the Wing Rib shown here. Each Wing Rib is held in place by a Rib End Ball that is attached to the end of each Rib Pocket. (See photo H for the Rib Ball Position.)

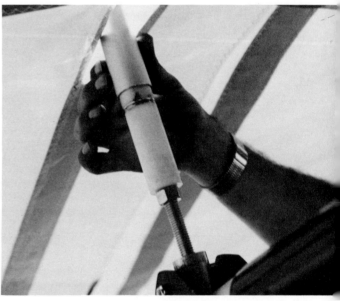

**J. THE KING POST TENSIONER** must be checked each Preflight. The Oversleeve slides to reveal the Locking Nut position. This is a vital but hidden component. If it is loose the Kingpost can fail and a wing collapse is a tragedy of epic proportions.

**K. THE KING POST TENSIONER NUT** must be up tight against the Tensioner, approximately 3" above its base, as here shown.

**N. THIS QUICK-SNAP IGNITION PLUG** is connected to the Off-On Kill Switch on the Throttle Assembly and to the Engine. It must be firmly seated to ensure continuous engine function.

**O. A PROPELLER FAILURE** is extremely dangerous. This Safety Pin holds the Hub Bolt Nut in place. If the Safety Pin is missing, the nut can work off the bolt and cause a serious propeller malfunction.

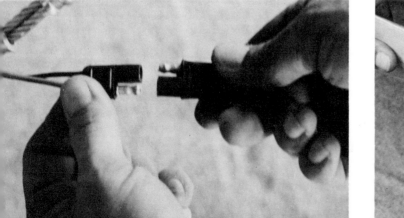

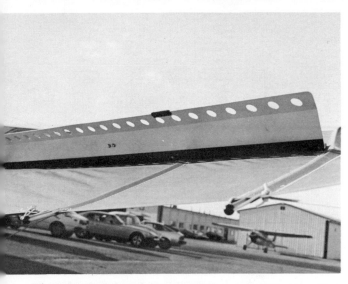

**H. SPOILERS SHOULD DEFLECT** upwards 90° without binding as the Control Stick is rotated side to side.

**I. RIGHT AND LEFT SPOILERS** are Velcroed to the Wing Surface approximately two inches above the Wing Seam. If too low, lift off the Spoilers and re-velcro down and re-attach the Spoiler Bracket to the Spoiler Cable.

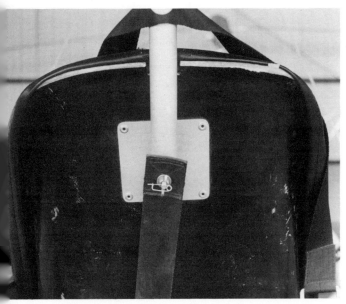

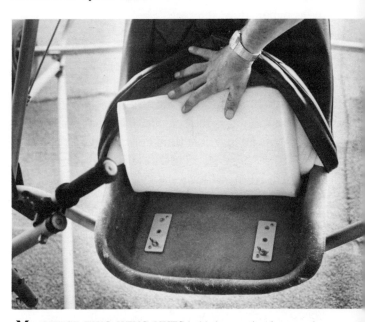

**L. THIS STRAP CONNECTS** the bottom of the Triangle Bar Base Frame to this attachment point on the back of the seat where the Verticle Rear Seat Tube holds the seat in place. Without this attachment the seat is structurely incomplete.

**M. THESE TWO WING NUTS** hold the seat in place on the bottom of the Triangle Bar Frame. The Sponge Rubber seat padding (under the seat cover) keeps the Wing Nuts from rotating off the bolts. They must be tight. The two Position Holes behind the Wing Nuts are for individual CG (center of gravity) adjustments.

**P. PREFLIGHT BEGINS AND ENDS AT THE NOSE.** Check the Tip Rudder Pedals for smooth operation. Also you observe the Nose Wheel Right and Left Steering action, Tire Inflation, Wheel Bearing condition and Wheel Lock Safety Pins.

## NOTE:

The EAGLE XL KEEL POCKET has a pouch sewn onto its side. It was put there to provide a place for the storage of the PREFLIGHT CHECK LIST CARD and LOG BOOK. You should use this card *every time* you PREFLIGHT the aircraft. To miss one item on the list is courting disaster. Remember, the more you know about all of the details of your airplane the better your piloting skills will become. A thorough, meticulous preflight is a mandatory procedure if you are to become a safe pilot.

# PILOT BRIEFING

## AERODYNAMICS

This briefing is to provide you with a background on the forces that influence flight in the Earth's Boundary Layer of Atmosphere and the terminology used when dealing with Aerodynamics and Meteorology.

Aerodynamic forces provide the Lift (L) necessary for flight and Meteorology makes flight possible or impossible.

To fly safely you must have an understanding of Meteorology and Aerodynamics and know what happens when they interact.

THE ULTRALIGHT PILOT'S PRIMARY CONCERN is the *stability* and *control* of an ultralight airplane as it flys in the atmosphere and the forces that have a dynamic effect on the aircraft while in flight.

Air in motion is the science of AERODYNAMICS. Air is an invisible fluid composed of a combination of gases called ATMOSPHERE that envelopes the Earth from sea level up to a height of about 30 miles (50 kilometers). Air also includes water vapor and various pollutants which can affect flight.

### EARTH'S ATMOSPHERE

THE EARTH'S BOUNDARY LAYER OF ATMOSPHERE has been described by astronauts flying in space as appearing very thin. Being able to see the Earth's atmosphere from space is caused by the cloud layers churning across the surface of the planet and the refractive qualities of the air as solar light penetrates through the air particles and refracts off their mass. Rainbows are caused by this refraction when sunlight reflects off the particles of moisture in the atmosphere.

The Earth's atmosphere blankets everything it touches all the way down in scale to the individual molecules that make up the total density (D) of the atmosphere. Without atmosphere, flying an airplane would be impossible. Flight is achieved by the interaction of the atmosphere and an airfoil moving through it.

Airplanes are supported in the air by aerodynamic forces acting upon the airfoil of the wing as it flys in the atmosphere.

Airplanes or any object moving in space, like the Earth (E), rotate about its center of gravity (CG).

All airplanes have a designed Flight Envelope that determines the airplanes VNE (never-to-exceed) Speed, Altitude and maximum gross weight. Should the airplane be flown beyond its designed VNE envelope, airframe failure is usually the result.

The airplane's Lift (L) and Thrust (T) forces created by the engine and propeller are designed to overpower the opposing forces of Drag (D) and Gravity (G) and in this action provide atmospheric flight for the airplane.

When flying, the pilot controls the airspeed, altitude and flight attitude or axis.

## AERODYNAMIC FORCES ON THE AIRPLANE IN FLIGHT

    AIRFOILS
    NEWTON'S LAWS
    BERNOULLI'S PRINCIPLE

AERODYNAMIC FORCES ARE:

    GRAVITY
    THRUST
    LIFT
    DRAG

AIRFOILS create lift by the action of air as it flows over and under the airfoil.

AIRFOILS are mainly found on the wing, tail surfaces and the propeller of an airplane. The Airfoil is a sectional element that is composed of specific parts.

AIRFOILS can be either Double surface or Single surface.

CHORD – The longest possible imaginary straight line connecting the leading and trailing edges of an Airfoil plane.

CHORD AXIS – The reference axis for the Geometric or Aerodynamic properties of an Airfoil.

GEOMETRIC CHORD – Used for Airfoil structural references.

MEAN AERODYNAMIC CHORD — Airfoil stability and balance is determined by this measurement and location.

LEADING EDGE is the forward point of the Airfoil that stabs (or bites) into the Relative Wind.

TRAILING EDGE is the aft point of the Airfoil.

UPPER CAMBER is the curved top of the Airfoil.

LOWER CAMBER is the bottom of the Airfoil.

## AIRFOIL PROFILE

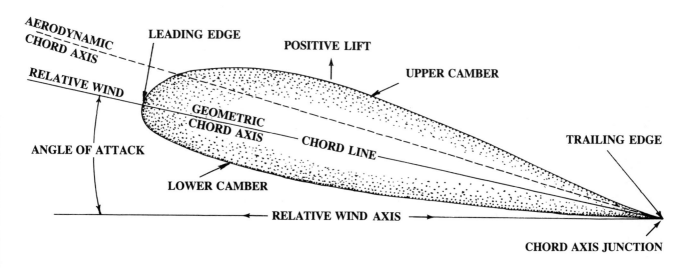

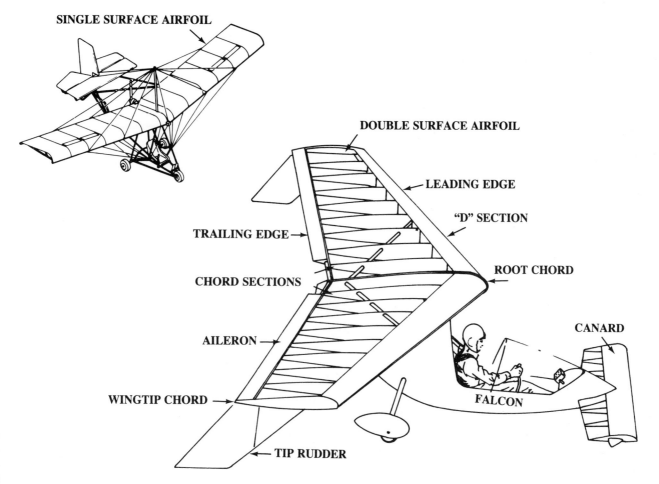

# PHYSICAL LAWS GOVERNING FLIGHT

## BERNOULLI'S PRINCIPLE

BERNOULLI'S Principle was discovered by a Swiss scientist named Daniel Bernoulli (1700-1782). His simplified principle states: Where the velocity of a fluid or gas is high, the pressure in the fluid or gas is low; and where the velocity is low, the pressure of the liquid or gas is high.

AIRFOIL design creates more lift upwards than down. The free air stream lines are closer together above the airfoil surface than below it. Where the air stream lines are close together, the air speed is faster than the air moving below the wing. Therefore, the air pressure below the wing is greater than at the top.

This lower pressure on the top of the airfoil allows the higher air pressure below the airfoil (sea-level pressure 14.7PSI) to exert a lifting force upwards where the air pressure is lower. Another force gives the airfoil life. It is the downward deflection of the free air stream as the airfoil moves forward into the relative wind. This downward deflection exerts a lifting force on the undersurface of the airfoil (Newton's third law of action-reaction). These two forces, Bernoulli's Principle and Newton's third law give the airfoil its lifting capabilities.

## SIR ISAAC NEWTON'S LAWS OF MOTION

### FIRST LAW OF MOTION

INERTIA—Every Mass (weight) remains in its state of rest or of uniform motion in a straight line (gravitational pull) unless it is forced (Vectored) to change that state of rest by forces (Vectors) applied upon it.

### SECOND LAW OF MOTION

ACCELERATION—The acceleration of a Mass (weight) is directly proportional to the Net Force Vectors acting on it. The direction of the acceleration is the direction of the applied Force Vector.

$$\text{Acceleration Vector} = a = \frac{\text{Force Vector}}{\text{Mass Vector}}$$

*The Mass (weight) of an object is multiplied by its acceleration and is equal to the applied Force Vector.*

### THIRD LAW OF MOTION

ACTION-REACTION—Whenever one Mass exerts a Force on a second Mass, the second Mass exerts an equal force on the first Mass. To every action there is an equal and opposite reaction.

## BERNOULLI'S PRINCIPLE

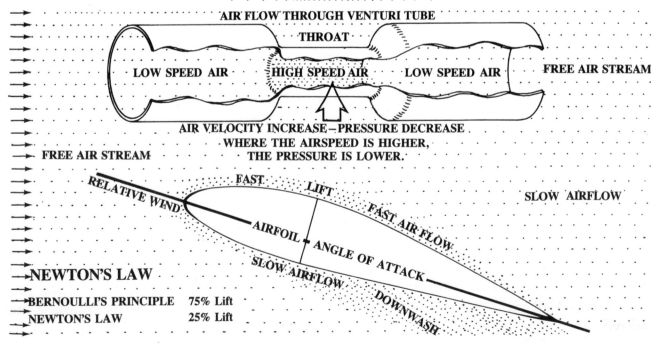

**EXAMPLE:** Hold a sheet of paper *below* your mouth and blow across the top of the paper. It will lift upwards because the air flowing over the top of the paper is moving faster than the air underneath the paper. The air pressure above is lower, so the air pressure (14.7 psi) below the paper pushes the paper up in the same manner as the airfoil rises as it moves through the air.

Hold a sheet of paper above your mouth and blow into the underside of the sheet of paper. It will lift upwards because the air molecules impacting with the bottom surface of the paper is a force. This force (Newton's Action-Reacton) causes the paper to rise.

Blow across the top — Bernoulli's Principle. Blow across the bottom — Newton's Third Law. These are the two forces that give an airfoil its lift.

# FORCES ON AN ULTRALIGHT IN FLIGHT

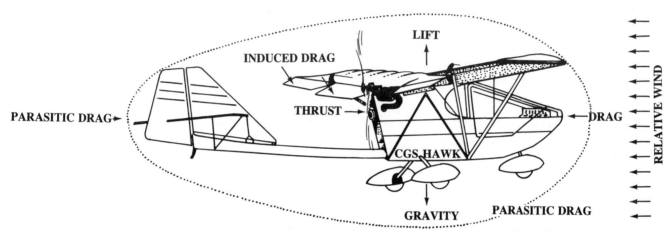

GRAVITY (G) is the force that holds all matter to the Earth. An airplane in flight has a gravitational force pulling at it at all times. Without LIFT, the airplane in flight would accelerate vertically toward the center of the Earth at 32.174 feet per second until it reaches a maximum speed of over one hundred MPH at impact.

THRUST (T) is a linear force that moves the airplane forward. This Thrust Force is created by the airfoil of a propeller rotating in the atmosphere by the power of an engine.

LIFT (L) is the aerodynamic force exerted by the air on an airfoil in a direction perpendicular to the direction of the moving airfoil. This interaction of the air (relative wind) against the airfoil causes the airplane to stay aloft. The hold GRAVITY has on the gross weight of the aircraft must be overcome by LIFT if the airplane is to fly.

DRAG (D) is the force exerted on the airplane which reduces its forward motion and is present in two forces: PARASITIC DRAG and INDUCED DRAG

PARASITIC DRAG is produced by the airplane moving through the atmosphere. Every part of the aircraft is involved. Wings, flying wires, pilot and pilot cage, landing gear, engine mass, tail, gaps in the wing are all PARASITIC DRAG components because they do not contribute to the development of LIFT.

INDUCED DRAG is produced as the airplane develops lift. Deflecting the ailerons down to gain lift, the angle of attack of the wing are components which produce INDUCED DRAG. As THRUST is increased, PARASITIC DRAG increases at a ratio of four-to-one while INDUCED DRAG decreases.

WING DESIGN Wings on the airplane is the component that permits flight. Flight is achieved by the interaction of the forces of air flowing over and under the WING.

The operational design of an airplane determines the (PLANFORM) of the wing. The PLANFORM of a wing is its shape as viewed from above and below.

# WING STRUCTURAL COMPONENTS

Wings are made of either wood, metal, fabric in single use or combinations of all the materials mentioned. Wings are designed to be strong, light and flexible.

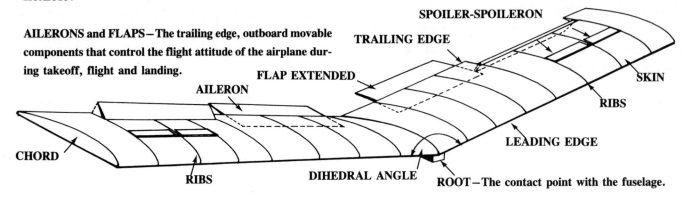

**AILERONS and FLAPS** — The trailing edge, outboard movable components that control the flight attitude of the airplane during takeoff, flight and landing.

**OUTBOARD END** — The part closest to the tip of the wing.

**SPAN** — The distance from wingtip to wingtip.

**WING TIP** — The outermost end of the wing.

**SPAR** — Extends the length of the wing from root to wingtip and supports the majority of the wing loads.

**INBOARD END** — The part of the wing that is closest to the fuselage.

**SKIN** — Material that covers the wing. Can be made of metal, fabric or wood.

LIFT TO DRAG RATIO (L/D). Increasing Lift while Decreasing Drag (L/D) is the most important factor in aircraft design and performance. An airplane with more Drag than Lift cannot fly. That is why aircraft have streamlined designs to reduce Drag. The L/D ratio is calculated with the engine off. The best glide slope angle and airspeed which determine the L/D of an ultralight is set by the manufacturer.

An airplane with an L/D RATIO of 12-to-1 means that the airplane has 12 times as much lift as drag. For example; an engine-off airplane with a 12-to-1 L/D RATIO, flying at 2,000 feet altitude Above Ground Level (AGL), will glide twelve thousand feet of horizontal distance for every thousand feet of lost altitude.

Every wing has an ASPECT RATIO. This ratio is determined by dividing the length of the span from wing-tip to wing-tip by the mean average chord (MAC) which is the measurement of the wing from leading edge to trailing edge. For example, a wingspan of 36 feet and a Mean Average Chord of 3 feet has an ASPECT RATIO of 12.

A long narrow, HIGH ASPECT RATIO wing creates more lift per square foot of wing area than a LOW ASPECT RATIO wing with the same square feet of wing area. For example; An engine-off airplane with a 5-to-1 ASPECT RATIO, flying at 2,000 feet AGL will glide 5,000 feet of horizontal distance for every thousand feet of lost altitude, even though its wing area is the same as the one with a 12-to-1 ASPECT RATIO.

WING LOADING To determine the wing loading of an airplane you divide the Gross Weight (Pilot, carry on load, accessories, fuel and the weight of the airplane) by the square feet of wing area. For example; The Eagle XL weight 253 lbs. without the pilot. Add the pilot's weight, say 160 pounds, and we have a gross weight of 413 pounds. Divide the gross weight, 413 pounds, by the total wing area, 177 sq. feet and we find the wing loading is 2.3. The higher the wing loading number the greater the power requirements are to fly the plane.

Ultralights, particularly the cable braced models have many Parasitic Drag components but the power plants and slow speeds made them very flyable. If the designed speeds are exceeded on any aircraft (VNE) never-exceed speed, the Parasitic Drag and G (gravity) forces encountered can completely destroy an airplane.

# RELATIVE WIND AND ANGLE OF ATTACK

The relative wind is the force of air flowing over the airfoil from the leading edge to the trailing edge and is always parallel to the airfoil's Mean Aerodynamic Chord, regardless of how the wind blows from meteorological forces. The speed of the airfoil moving through the air determines the Relative Wind's velocity and lifting force of the airfoil.

The degrees of the angle, from the horizontal plane to the relative wind's flow direction into the leading edge of the aerodynamic chord line of the airfoil determines the Angle of Attack.

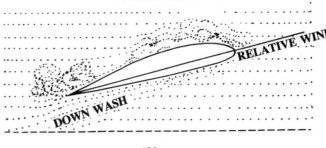

18°

**4. HIGHER ANGLE OF ATTACK.** Airfoil turbulence increases. Lift continues but drag is also increasing.

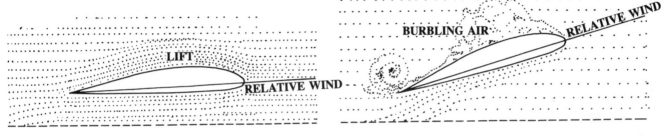

4°

**1. LOW ANGLE OF ATTACK** into the relative wind. Lift normal. Airflow smooth.

20°

**5. CRITICAL ANGLE OF ATTACK** is rapidly approaching. The airfoil will stall unless more power is applied to Thrust.

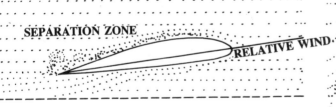

10°

**2. INCREASE ANGLE OF ATTACK.** More lift. Airflow smooth.

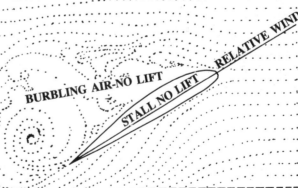

**6. CRITICAL ANGLE OF ATTACK IS REACHED.** The airfoil stalls as the air flow over the wing breaks up into severe turbulence. The nose of the Ultralight pitched down and starts to descend until the Relative Wind again flows smoothly over the airfoil.

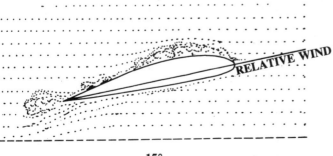

15°

**3. ANGLE OF ATTACK INCREASED.** More lift but Drag is present with turbulence over the trailing edge of the airfoil.

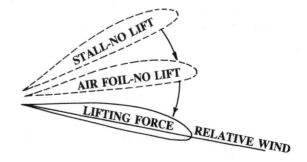

# FLIGHT SIMULATOR – UNPOWERED

Kris Williams was all business as he began my instruction.

"Put your helmet on and hop into the seat, and we'll begin with the flight simulator."

I slipped the helmet on, fussed with the chin strap connections—I seemed all thumbs. Bob swiftly helped me get the fastening closed.

Climbing into the cockpit was a momentous occasion. I gingerly snaked through all of the flying wires and restraining cables that seemed to be all over the simulator. After disentangling my feet, which had been trapped in a web of cables, I settled myself into the bucket seat and searched for the safety belt. Again four hands assisted me. Shoulder harness on, chest straps velcroed down to keep them there, lap safety belt secure and tight and a final deep breath. I looked up as Kris asked, "Feel comfortable?"

I smiled and nodded affirmatively. I felt like a lion in a cage: Roaring to go but strapped hard in the Eagle.

**YOU'LL FEEL APPREHENSIVE** when first learning to pilot an aircraft, but the tension soon dissolves into focused concentration as you learn to feel the effect of each of the controls on the flying aircraft. CAUTION: ALWAYS WEAR YOUR HELMET AND SHOULDER AND CROSS-CHEST SAFETY HARNESS RESTRAINERS WHEN ABOARD THE PLANE. The right hand is on the control stick, and the left is on the throttle.

To go up—pull back on the stick; to go down—push forward. To make the wing tips go up and down, move the stick side to side. To swing right and left alternately, press your toes down on the foot pedals!

**HERE IS THE EAGLE XL FLIGHT SIMILATOR.** The student is in the cockpit, and the instructor crouches right in front of him, so that he can talk directly to the student, you. Restraining cables loosely tether the Eagle XL to the platform pushed by the station wagon. This allows you to operate the controls while being propelled down the runway at the Eagle's flying speed of twenty-six. You learn takeoff rotations, straight and level flight, right and left rolls into banks and yaws, time lag awareness between control manipulation and aircraft response to the control movement, flares and landings. You learn the relationship between control movement and aircraft response; you use both feet to activate the wingtip rudders, and the right hand for both spoileron operation and up-and-down rotation of the canard.

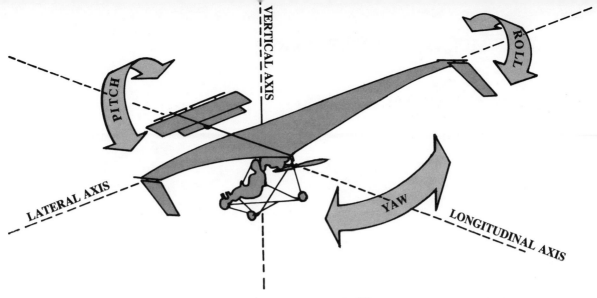

**EAGLE XL: 3-AXIS CONTROL**

**PITCH:** The up-and-down nosing of an airplane or spacecraft about its *lateral* (wingtip to wingtip) *axis*. On the Eagle XL, PITCH CONTROL is done by rotating the CANARD wing UP and DOWN.

**ROLL:** The rotation of an airplane around its *longitudinal* (head-to-tail) *axis*. On the Eagle XL, ROLL CONTROL is done by raising or lowering the SPOILERONS, which are located near the wingtips on the top center of the wing. When the right spoileron is up, the drag on that side rolls the wingtip down, while the opposite wingtip necessarily goes up; with the left spoileron up, the reverse happens, with it rolling down and the other wingtip going up.

**YAW:** The horizontal rotation of an airplane about its *vertical* (perpendicular to the lateral and longitudinal) *axis*. On the Eagle XL, YAW CONTROL is done by rotating the WINGTIP RUDDERS about their vertical pivot axes. Inboard deflection of the right wingtip rudder YAWS the Eagle XL to the right; inboard deflection of the left wingtip rudder YAWS it to the left.

"What we're going to do now, Rick, is drive down the runway, and when we gain enough speed, I'll tell you what to do. The Eagle's engine will not be running, so you'll hear everything I have to say. I will be right here in front of you as you operate the controls of the Eagle."

My heart was clicking along at a ten percent increase. Hands with damp palms were kept still. They wanted to fidget and fuss with straps, helmet, anything. I was nervous and extremely excited at the same time. Kris motioned to Bob and the station wagon started down the runway. The air began to blow across my cheeks. I could feel the Eagle begin to lurch into the freshening wind.

"Notice, Rick, how the wings fill with air and the cables tighten as we gain speed."

I looked up at the wings. The fabric was taut. The bird was trembling.

"Now put your hand on the control stick and keep your feet on the tip rudder pedals. We're going to go a little faster."

My jaw tightens as Bob aims the wagon down the runway with more gas.

"As we gain speed, Rick, rotate slowly back on the control stick."

The wind was brisk; I followed instructions. Back came the stick and the Eagle lurched hard against the restraining cables with the snap and rattle of shackles. The Eagle was flying, but the hands of the fledgling were much too strong. I moved the stick what seemed a little, and the Eagle bucked left, then right, straining to release from the wires that held her down.

"That's OK, Rick, you're controlling all the actions. Be a little lighter with your movements on the stick. Try not to overcontrol, be gentle."

Bob decelerated and the Eagle thumped down on the simulator platform as we reached the end of the runway. Bob turned around and lined up to make another run. We accelerated down the runway. The wind filled the wings as Kris leaned close to me.

"Rotate up slowly this time, Rick. Make it smooth. Pull back on the control stick."

I'm going to do it right this time, I thought. Slowly I pulled the stick back. The Eagle bucked and snapped against the restraining cables with a twang. I held the bird up tight against the cables, then gently pushed the stick forward. The cables slackened, but the Eagle lurched steeply to my right, with a rattle of cables and shackles straining against the restraining attachments. Kris spoke.

# FLIGHT SIMULATOR—UNPOWERED

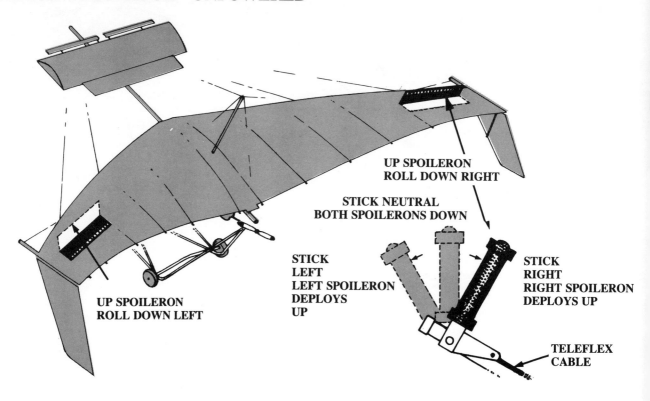

## ROLL CONTROL BY ROTATION OF THE SPOILERONS

**WITH THE CONTROL STICK IN NEUTRAL,** both right and left spoilerons remain lying flat on the upper surface of the wing. Move the STICK RIGHT and the right spoileron rotates up 90°, increasing the drag and making the right wing drop into a ROLL. Move STICK LEFT, the left spoileron rises 90°, the right spoileron drops back flat against the top of the wing, and the left wing ROLLS down.

Kris said, "You depressed the right foot pedal and deflected the right tip rudder. That's what brought the right wing down. Release the pressure on the right foot pedal and the plane will level."

I did and the wings levelled with the horizon—which I noticed for the first time.

"That's good, Rick. Hold it there and make the Eagle fly without straining on the simulator."

Deep concentration. Carefully I centered the stick. The Eagle's wings wobbled but remained level. I slowly eased the stick forward. The cables slackened. Stick rotation forward a little more. The Eagle sat down on the deck of the simulator. Back stick and up she went. This time I caught it before that teeth-gritting snap and bang that happened when I pulled back too far. A quick glance at the restraining cables. They were slack. The Eagle was flying.

"Give the stick a slight move to the right. That will lift the right spoileron and make the Eagle bank to the right. Easy now."

My right hand bent to the right, just a little, as Kris had instructed. The Eagle made a slight tilt to the right, and then kept going down. The left restraining cable drew taut as the right rear landing wheel thumped down on the simulator. I swung the stick to the left. The right wing rose up sharply, and when the wing was level with the horizon I brought the stick back to center—but the right wing kept on lifting up. The restraining cables twanged again, then went slack as the Eagle flew. Bob braked to a stop on the end of the runway and the Eagle thumped down. I was sweating profusely

Kris stood in front of the simulator and asked, "Did you notice something different this time, Rick?"

I cleared my head of the spin raging inside and collected my thoughts.

"I kept the Eagle flying without slamming against the restrainers, but when I started into the banks, she seemed to have her own mind. I brought the stick

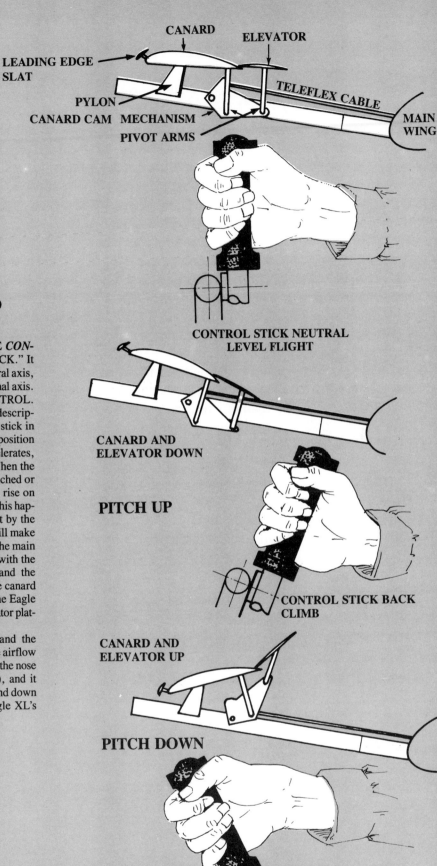

## PITCH CONTROL BY ROTATION OF THE ARTICULATING CANARD

**YOUR RIGHT HAND OPERATES THE *CONTROL* STICK**, usually just called the "STICK." It controls the PITCH or rotation about the lateral axis, and the ROLL or rotation about the longitudinal axis. Here we shall only deal with PITCH CONTROL. (See the aerodynamics chapter for a fuller description of these terms and functions.) With the stick in Neutral, the canard remains in neutral flight position parallel to the main wing. As the aircraft accelerates, the airflow billows the main wing fabric. When the [relative air-] speed of twenty-six mph is reached or slightly exceeded, the Eagle XL will gently rise on its own, without any control manipulation. This happens because the angle of the canard, preset by the manufacturer, is angled up. This up angle will make the nose of the Eagle XL slowly rotate up. The main wing follows this lead and flies right along with the canard. Pull (rotate) the STICK BACK and the Teleflex cable angles the trailing edge of the canard DOWN, increasing its LIFT. The nose of the Eagle XL rises, (PITCH UP) as it lifts off the simulator platform and is airborne.

Push (rotate) the STICK FORWARD and the canard and elevator extend up, deflecting the airflow from the top of the canard. This action forces the nose of the Eagle XL to drop (PITCH DOWN), and it lands on the platform. After a few trips up and down the runway, you will have learned the Eagle XL's response to its flight controls.

53

# FLIGHT SIMULATOR — UNPOWERED

back to center on each bank, but the Eagle kept right on going up."

"That's right. There is a slight time lag before the Eagle reacts. That lag is very important and you must compensate for that lag in each executed move of the controls. You held the control in the moment position until the Eagle responded and performed the maneuver. Then you initiated the correction procedure. By that time it was too late. The plane kept right on flying into the previous instruction you gave it. This is because of that time lag. On the next run down the runway remember this, and master that *time lag*. It will make you a safe flyer when you get it right."

Three more runs up and back on the runway and I had that time lag under control. I took the Eagle up, held her level, banked to the right, then wings level, then bank left, wings back to level, nose up, nose down, and a quiet return to the carpeted simulator without that annoying thump, bang and rattle of the cables.

"That's excellent, Rick. Now on this run I want you to get the feel of the flair before you set the plane down."

Back down the runway at a fast clip. Stick back and the Eagle flew. I worked on the banks and turns until they became smooth and silky. Then Kris asked me to flair. I waited until Bob began to decelerate, and I pulled back on the stick. The nose of the Eagle rotated up into the air, the rear wheels touched down on the simulator's deck, then the nose wheel came to rest as Bob stopped. Kris jumped off the simulator.

"You did fine, Rick. We'll take a break and then begin the towing lessons."

I clambered out of the web of simulator wires, took off the helmet, wiped my damp forehead with a red bandana my friend Lynn had given me, and sat down. I was exhausted. I felt like I had worked several hours at hard labor, when in fact we were on the runway hardly an hour.

"They sure put you through the paces fast here. I'm ready for a long break," I thought, as we drove back to the Training Center.

Within minutes the reality of what had just happened hit me. I'd flown! — but I had been so busy with the immediate problems of flying that the full impact of the experience had evaded my senses. Now I felt it, and energy began to flow through me. I wanted to fly that Eagle again!

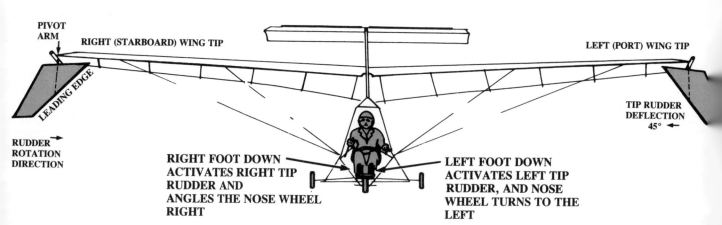

## YAW CONTROL BY ROTATION OF THE WINGTIP RUDDERS

**YAW IS CONTROLLED BY THE FOOT PEDALS.** DEPRESS RIGHT FOOT PEDAL, the right wingtip rudder pivots up to 45° inboard about its vertical axis. The resultant drag makes the nose YAW to the right. DEPRESS LEFT FOOT PEDAL, simultaneously letting up on the right, and the left wingtip rudder can now be deflected inboard similarly, with the right wingtip rudder returning to neutral. The resultant drag on the left wingtip rudder causes the Eagle to YAW to the left. DEPRESS BOTH FOOT PEDALS simultaneously, both wingtip rudders are deflected inboard, the Eagle XL loses airspeed, and the nose drops.

### NOSEWHEEL CONTROL

The foot pedals of the EAGLE XL also control the nosewheel steering.

54

# STUDENT-PILOT PREPARATION

## PHYSICAL CONDITIONING

★ You may find it difficult, but get plenty of sleep the night before you begin your flight training.

★ Do not drink alcohol or take any drugs (unless under physician's orders and your flight instructor is fully informed) twenty-four hours before or during flight training. To do so can affect your mind and physical capabilities and will negatively alter your coordination of the flight controls and judgments. The physical and mental stresses placed on you during these first lessons are considerably more than at any other time during your flight training. So you must be physically and mentally prepared. Five hours of *hands-on* pilot training on these first few days require more stamina and clear thinking than a calm, windless day of actual flight.

★ Each training day get up early enough to eat a leisurely, healthy, breakfast. A gnawing stomach from lack of or bolted down food or jittery nerves from too much coffee on an empty stomach are very distracting. Once you are out on the flight line and harnessed tight into the cockpit there is absolutely no time for snacks or drinks of any kind. Even going to the bathroom is difficult during this phase of the training.

★ Your body must have all the available energy it can muster if you intend to concentrate and handle your aircraft to the best of your abilities.

★ Airports and Ultralight flight parks are by their very nature out in the open. In the summer they are hot and temperatures frequently float in the high nineties or hundreds. In the winter, thermometer readings in the low thirties or twenties are not uncommon. So, your body and mind must be in top condition in order to handle the extremes of flight training.

*Few things are brought to a successful issue by impetuous desire, but most by calm and prudent forethought.*
THUCYDIDES

## CONCENTRATION— GOOD HABITS

★ To fly well you must have 100% concentration and good habits. Old habits (which in flying are usually more bad than good) must be ruthlessly eliminated. *Preflighting* your aircraft is the first good habit you must develop and perfect. Each item on the preflight check list (and as you progress, many others) must be read, thought about, looked at, felt with your hands and a decision made as to the airworthiness of the item observed before you move on to the next item to read and check on the preflight card. Good habit demands that this procedure be followed without *any distractions whatsoever*. If one item is overlooked, *a bad habit has emerged* in your preflight procedure and, that one item, overlooked, may one day cause an accident. Your enthusiasm to fly can be your greatest friend or worst enemy. It rests solely on the degree of responsibility you, the pilot, assign to your newly acquired habits.

★ As each day's flight training finishes and you retire, review all that has happened in your training that day and make notes, mentally and on paper, of what you have learned and what habits you have that are good, what habits you have that are bad and then proceed to eliminate the bad through truthful, objective introspection and self-criticism.

*Since habits become power, make them work for you and not against you.*
E. STANLEY JONES

## TEMPERAMENT

★ Everyone should express emotion. Overreaction is the enemy. Flying requires all your attention. You must leave those emotions (that lead some into kicking tires on the ground or slamming fists into the wall) on the ground before you fly. Many people crumple when they make a mistake and are criticized by the instructor, and internal anger fogs their judgment. Then on the next lesson their concentration is dislodged by that anger and the student pilot ground-loops or makes some other serious mistake, all because of uncontrolled anger or emotion.

★ The good pilots have learned to take errors or mistakes in stride by being cool, calm and collected and correct those errors and mistakes before they grow into demons and cause a serious accident.

*What we think, or what we know, or what we believe is, in the end, of little consequence. The only consequence is what we do.*
JOHN RUSKIN

## PEER PRESSURE AND PRIDE

★ One of the greatest hazards a pilot can face is forcing him or herself into flying beyond their capabilities pilots with thousands of hours of logged flight time. These errors in flying judgment usually occur when there are other pilots standing about, talking of flight, and someone challenges you into doing something in the air you are not fully qualified to do and you do it because of Peer Pressure and Pride.

★ The pages of the ultralight magazines and other flying journals always carry stories about fatal crashes caused by Peer Pressure or Pride or both.

★ When the pilot's pride, influenced by ego, enters into a decision to fly because of peer pressure you have the ingredients of tragedy.

★ You do not allow PEER PRESSURE or PRIDE to influence your decision to fly, if you are to survive.

*Man's last freedom is his freedom to choose how he will react in any given situation.*
VICTOR FRANKEL

# TAXI

**THE WINGTIP RUDDER PEDALS** turn the nosewheel of the Eagle XL. Its turning radius is wide, and turns must be anticipated and initiated with proper pedal control well in advance. On sharp turns, 90° or more, you can either use your feet on the runway surface to edge the nose of the Eagle into the desired heading, or dismount, lift up the nose, and walk it around.

CAUTION: NEVER EXIT THE EAGLE WHILE THE ENGINE IS RUNNING. To do so can lead to serious accidents. If in doubt while taxiing, FLIP THE KILL SWITCH OFF and stop the engine.

In the above photo, the left foot pedal is depressed and the left wingtip rudder is deflected inboard 45° about its vertical pivot arm, and the nosewheel is being swung to the left.

**DAY 1**

The Eagle XL was quickly and efficiently relieved of all the restraining cables that attached it to the Simulator platform by Bob and Kris. I assisted them during the process, familiarizing myself with the procedure.

We carefully lifted the Eagle from the platform and placed it on the tarmac, and I began a thorough preflight under the watchful eye of Kris. He remained silent while I concentrated on this vitally important task. His main concern was that I did not miss anything.

"I've completed my preflight. Did I miss anything?"

"No."

Kris walked up to me and handed me a helmet as Bob fired up the Cuyuna engine and set it on Idle, then took a position over me with both hands on the triangle cockpit cage as Kris spoke.

"This is your taxi practice. You're going to follow us behind the station wagon up to the taxi training runway. Use the throttle to keep yourself moving a constant distance behind us, and remember the nose wheel steering is controlled by your foot pedals. Push right and the nose wheel turns right; left foot press and the nose wheel turns left. Use the brake post there between the two foot pedals to slow down or stop. You will notice that the brake is a piece of flat metal with the post attached to the center. The brake acts on friction of the metal plate pressing against the rear of the tire. Hold your foot on the brake post and we'll run the engine up and you can feel how much pressure you need to apply to hold the plane from going forward."

Kris ran up the throttle and as the Eagle prop whirled I could feel the surge forward. I pressed hard and we stood still, but to hold the Eagle back I had to push my heel down so hard my back was pressed up tight on the back of the cockpit seat. Kris set the throttle back to idle. I kept my foot on the brake with less pressure than before, but still firmly planted.

"Are you reading me, Rick?" Kris held the hand radio to his lips.

"Rotate the canard if you are." I nodded and worked the canard up and down.

"Good. I'll be in the back of the station wagon, well ahead. Stay right behind us and if anything happens and you want to stop, just use the kill switch. You know where it is?"

I lowered my left hand and put my finger on the

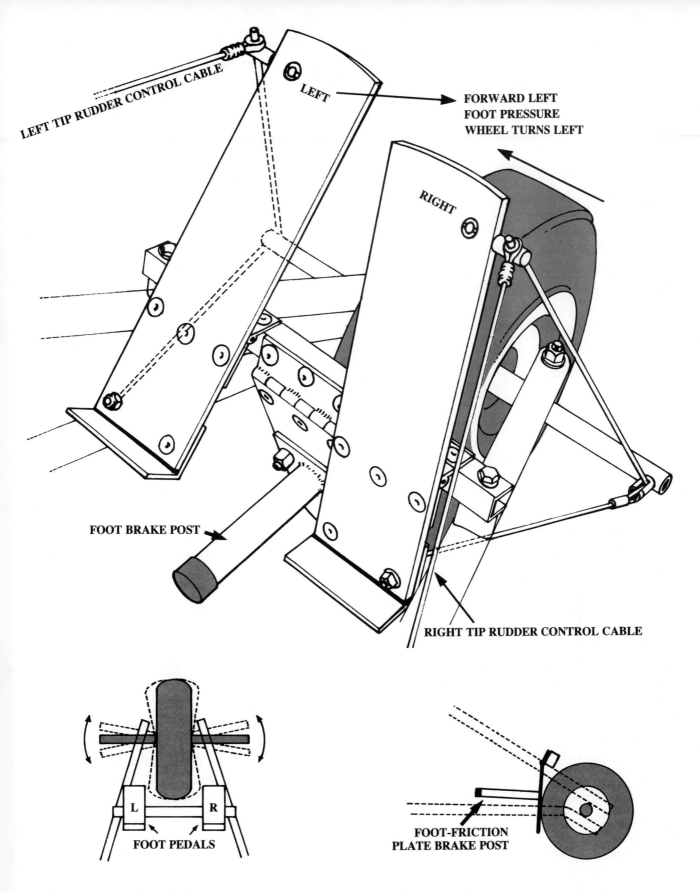

**PUSH DOWN THE LEFT FOOT PEDAL** and the nosewheel turns left; push down the right and it turns right. The steering action on the Eagle XL is hard, and requires heavy foot pressure to effectuate turns. Practice solves the problem, but be ready in the beginning to make emergency stops when it gets hard to turn the nosewheel. In wet weather, the friction plate of the nosewheel brake may slip; if necessary, use both feet on the nosewheel brake post to stop the aircraft.

CAUTION: NEVER PUT YOUR FEET DOWN ON THE GROUND WHEN THE ENGINE IS RUNNING: HIT THE KILL SWITCH FIRST. Otherwise, the forward-moving aircraft will sweep your feet back under your seat and seriously injure you.

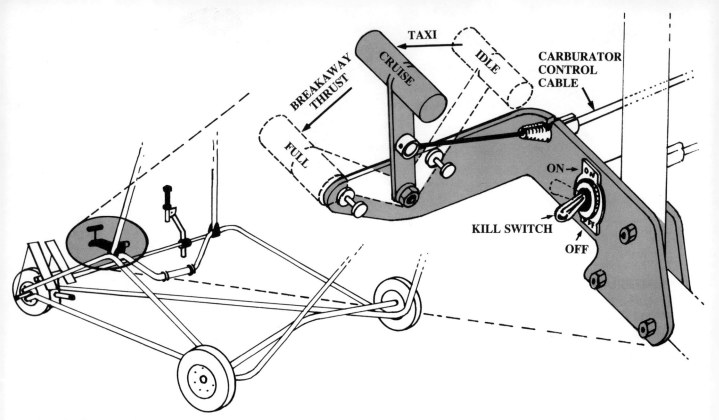

## THROTTLE CONTROL

**BREAKAWAY THRUST** is needed to move the aircraft forward. Once forward motion is obtained, set the throttle position to keep the aircraft at a safe taxiing speed. Fledgling pilots tend to overcontrol the throttle by advancing the *Throttle Control Lever* too far forward, into the FULL THROTTLE position. This causes the aircraft to surge forward at an alarming speed. To stop forward motion, the novice first uses the brake, and then retards the throttle all the way back to IDLE. The braking sequence may be reversed. Throttle procedures become smooth and even through practice. The skillful simultaneous use of **THROTTLE** and **BRAKE CONTROLS** and **NOSEWHEEL STEERING** technique is required for safe aircraft taxiing procedures.

Kill Switch and nodded affirmatively.

"OK, Rick, now follow us, and keep in the center of the taxiway."

The first problem I had to lick was APPREHENSION. My hand was on the throttle. My mind talked to me. OK, Rick, move it out. Slowly and gingerly I advanced the throttle and the Eagle crawled slowly forward. We were on an upgrade. A little more throttle; I pushed it too far ahead. The Eagle surged ahead and immediately I was faced with steering the Eagle. When we lurched forward I must have had pedal pressure on my left foot, from the habit of driving a car with the foot on the brake pedal. The Eagle started sweeping into a left turn, well off the center of the taxiway. I compensated with my right foot pressure on the nose wheel pedal. The bird swung to the right and I started using both feet on the steering pedals — but now we were going too fast. I'd been so absorbed with the foot pedal steering, keeping my eyes glued to the taxiway, that I was fast approaching the rear of the station wagon. If I didn't slow down, I'd run right into them. Kris's voice spoke, "Back on the throttle, Rick, and use brake pressure. You're do-

ing fine. Keep the nose wheel on the yellow line ahead as we turn towards the flight line and try to keep an even distance between us."

I pulled the throttle back to idle, and slowed the Eagle down with foot pressure on the brake post. But now a new problem arose. My left foot responded with my right foot and applied pressure when I pushed my right foot hard on the brake to stop. The nose wheel angled and we turned sharply left. A hangar door loomed ahead of us. I was five feet from the yellow line and heading for a left wing tip collision with the hangar door. Foot off the brake pedal. Hard right foot pressure and we swing to the right, the wing tip missing by a safe, but small, margin. I aim the front nose wheel towards the yellow line, and advance the throttle very slowly, and then bring it back just a touch, to keep us moving at a steady forward pace behind the station wagon.

"This is tricky stuff," I thought as I concentrated on coordinating my feet with the steering and braking and my erratic control of the throttle. I did a lot of prop rpm's and brake control. Too jerky for my peace of mind — and another obstacle now faced me.

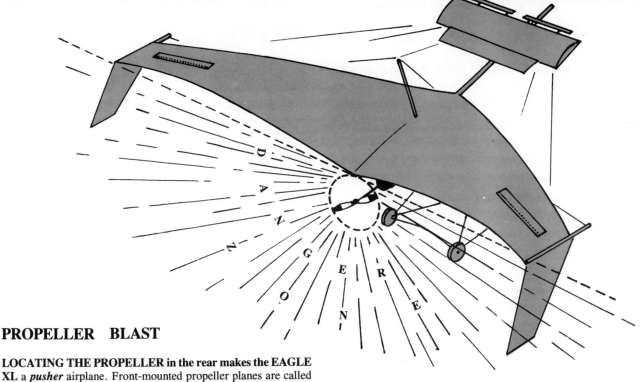

## PROPELLER BLAST

**LOCATING THE PROPELLER in the rear makes the EAGLE XL** a *pusher* airplane. Front-mounted propeller planes are called *tractor* aircraft. When taxiing on a rocky surface or dirt, the dust and debris blown to the sides and rear by the **Prop Blast** can cause considerable damage. To prevent Prop Blast accidents while taxiing, always carefully observe the sides and rear, as well as forward.

The station wagon had reached the end of the taxiway between the hangars and was making a ninety degree left turn up a slight incline. The yellow line followed their course.

As we started into the turn I gave it left pedal pressure. The Eagle started around, but not far enough. The yellow line drifted away from us to our left. More left pedal pressure and more throttle—we were stopped on the incline. Too much throttle again. The Eagle scampered up the incline as I centered the nose wheel on the line and for the first time I used my left foot on the brake post, while steering with my right. Throttle back and both feet in control of our direction behind the station wagon. The bird was moving along at the right speed and direction. Kris's voice came through.

"We're coming up on a runway ahead. We'll stop there before crossing. On taxiways ALWAYS STOP AND LOOK *BOTH* WAYS to see if there are any aircraft taking off, landing or taxiing in your way. Make certain the runway is clear, then go ahead and expedite the crossing. No lingering." They stopped and then drove across. I stopped and looked both ways: Nothing landing or taking off. I ran up the throttle hard and dashed across.

Down another access way to the end of the taxiway and they stopped, got out and walked up to the Eagle as I hit the kill switch and came to a stop.

## TAXI GROUND OPERATIONS

TAXI PREPLANNING: Observe the surface conditions for the Taxi Route: Wet or dry, paved, dirt, rocks, grass, open or plowed field; obstructions such as buildings, runway markers or edge lights and posts, parked cars or aircraft, trees, walls, etc. Remember that the Eagle XL's wingspan is thirty-five feet, so make mental note of all obstacles and contingencies that may be present before you taxi the aircraft.

GROUND WINDS can cause serious problems when you are taxiing. Remember, the Eagle XL is designed to fly, and wind conditions on the ground can have the same effect on the aircraft as those in the air. If the wind is blowing more than eight-to-ten miles per hour, tie down the aircraft and wait for the better conditions of lower wind speeds. The best winds for taxiing ultralights are no more than a mild three-to-five mph, and blow from either directly ahead or directly behind. CROSSWINDS must be carefully studied, and often avoided entirely. A CROSSWIND GUST can cause a GROUND LOOP by blowing under one wing, flipping the aircraft over, and ramming the opposite wing into the ground. Worse yet, it can flip the entire aircraft into the air and over, slamming it back into the ground and causing considerable damage and injury to the plane and pilot.

## UPWIND AND DOWNWIND TAXI TURNS

Make turns extremely carefully into the UPWIND or DOWNWIND direction, otherwise you may GROUND LOOP as you enter the turn. Sometimes it happens so quickly that you have little or no time to compensate for the lethal crosswind. Always note the wind conditions, and if they seem too rough, kill the engine, dismount, and tie the aircraft down, or have someone hold the wing as you walk the plane back to the hangar.

## CROSSING ACTIVE RUNWAYS

Always stop and look both ways to make certain no aircraft are taxiing, taking off, or landing in your path. Active aircraft always have Right-of-Way. After making sure the runway is clear, make your crossing quickly, but make all turns gently.

# UNPOWERED TOW TRAINING
## NOSEWHEEL STEERING ACROSS RUNWAY CENTERLINE

I glanced at my watch. It was 11 a.m. I was about to begin my first runway lesson. The taxiing experience still had me going. My mind was filled with images of the XL's 35ft. wings bobbing up and down; the roar and thrust of the engine; that nose wheel and brake operation plagued my memory. "I will master that nose wheel," I thought.

I was examining the pivot attachment on the left tip rudder when Kris walked up to me. He spoke as he adjusted the chin strap on his helmet. "Before each training session I take the aircraft up for a flight. I do this so I know how it flys and to check out wind conditions."

I watch Kris with the eyes of an eagle as he did his preflight, making mental notes on his technique. He was fast, thorough and all business. I noticed the firm way he ran his hands along the leading edge of the wing. Tugging on the flying wires then angling his head as he peered sharply up into the fabric access ports at critical junctions on the wing. Checking the spoilerons, his fingers quickly moved over the Cuyuna engine with the precision of a concert pianist. When he had finished he slipped into the cockpit, snapped on the safety harness, and shouted "Clear Prop." He fired the engine and began his takeoff roll. I counted the seconds. One. Two. Three. Four. Kris pulled the stick back and the Eagle smartly nosed into the air.

**KRIS WILLIAMS FLYING** the training Eagle XL. This first flight is a ritual he does each morning before student training sessions in the Eagle XL ultralight.

**KRIS ROLLS THE EAGLE XL** into a turn. INSERT: Touch down. Notice how the crosswind holds the left (port) main wheel off the runway as the right (starboard) wheel makes contact.

Bob walked up to me and drew my eyes from the beautiful sight of Kris drilling the Eagle XL into the eastern sky. His voice calmed my excitement and brought me back to earth.

"Your taxi this morning was erratic. Try not to use the throttle excessively. Find a setting that will move the aircraft forward at an even pace. You kept advancing the throttle then pulling it back to idle. You are not alone on this problem. Many new students do the same thing. It takes practice. Remember to keep the throttle at a setting that keeps the aircraft moving forward at a constant, manageable speed."

I went to the point that was bothering me.

"I had a tough time with that nosewheel steering and the coordination of the brake pedal and the throttle. I was all over the taxiway." I laughed. This morning reminded me of the day when I was a kid on my first tricycle ride. I drove that bike right into a thorn hedge and came out bloody and sore. The mental image of that first bike ride influenced my actions and took my mind from the immediate job at hand. "I'll concentrate and get that steering down."

Bob smiled. "You're doing fine, Rick. Remember, the nose wheel steering on the XL is the same as conventional aircraft. Push right, go right. Push left, go left. A few more taxi runs and it will come to you. Every pilot must master the taxi. Ground handling is the tough part of flying."

I laughed, remembering a news story about one of the Near East kings who piloted his own jet airliner and while he was taxiing up to his parking berth at LaGuardia airport, ran a wingtip into a parked jetliner. The picture in the paper showed a twisted dangling wingtip and an embarrassed king. No doubt he had trouble with the nose wheel steering.

Bob put his hand on my shoulder. "The king is not alone. Taxiing takes a lot of patience and practice. Just watch that throttle control."

Both of us stopped talking as we watched Kris roll into a 45° bank, line up with the runway, pitched the nose down, did his round-out, flaired and set the mains down with a soft kiss on the tarmac. He smartly taxied up, did a 180° turn so that the nosewheel pointed straight down the runway, hit the kill switch, un-buckled the seat belt, hopped out and walked up to us. He pointed to the windsock as he spoke. "There is a slight cross-wind blowing in from the North West. It is not serious enough to keep us from our ground exercises this morning so, we'll hook-up the tow lines and give you some slow taxi lessons, Rick." He handed me my helmet and began to attach the tow lines to the Eagle XL.

Within minutes I was strapped into the cockpit and listened as Kris outlined the lesson.

"What we are going to do now is tow you down the runway at a slow speed. I want you to keep the nose wheel right on the white centerline. If you drift off to one side or other bring the nose wheel back on the line. Any questions?" I shook my head as I said, "No."

"Good. I have the radio connected and I'll talk to you as we roll down the runway." Kris hurried to the tailgate of the station wagon as I thought to myself. "Concentrate, Rick. Get it right and make it smooth."

Kris spoke over the radio. "If you are receiving me, rotate the canard up and down." I did. "Good," he said; the tow line tightened and we started down the runway.

I had my eyes glued to the white center dashes slowly moving under me. The Eagle started to move to the left. Right toe pressure and it swung to the right. "Too far over," I thought. Left foot pressure on the rudder pedal. Over to the left, then right pressure. Over right then a little left pedal and we were on the centerline. I held it there, swaying slightly left and right when Kris released the tow line and I rolled to a stop. He dashed up to me and swung the nose of the XL to a heading back down the runway. "You did fine, Rick. Try keeping the nose wheel right on the centerline. You crossed over it several times. Right on the line now."

Down the runway again. A little faster this time. The nose wheel tracking perfectly, then suddenly the Eagle started tracking off to the right. I gave it left pedal but she seemed to stay on a right tack. More left pedal. The XL responded but not enough. Hard down on the left pedal. Slowly the bird started to move back to the line. I held the pedal down hard until we were on the line. Centered, I let up on the pedal pressure and the XL started back to the right. "What's going on," I thought as I rammed my foot down hard on the left pedal. The XL switched left and back on the line. I kept it there by holding a steady left foot pressure on the left pedal as Kris released the tow line and I rolled to stop. Kris was at my side and as he swung the Eagle's nose back down the runway, he asked.

"Notice anything different that time, Rick?"

"It kept swinging to the right as we headed down the runway. It took a hard left pedal to keep it centered." Kris pointed to the wind sock. "It's that cross wind. Compensate for it on the next run down the runway."

The line went taut and we hustled ahead. It took constant left pedal pressure to keep it on the line. but I was unhappy with the run. The Eagle kept a slight oscillation right and left as we moved forward. It bothered me. I asked Kris at the end of the run.

"I had a lot of difficulty holding the ship on the line. She kept swinging right and left. I compensated for the crosswind but the oscillations continued."

"Watch your foot pressure on the pedals. Don't over-compensate. We'll go a little faster this time. Hold it on the line."

Down we went. Total concentration. Left pressure to hold that crosswind down and a little right on and off to hold her on the line. Suddenly I had it together. We were tracking straight down the centerline without oscillations. Kris released and came up to me.

"Notice anything different this time?"

"Yes. The oscillations stopped and she went straight down the line. It seemed a little easier this time."

"What did you do that was different?"

I shook my head. "I concentrated on the nose wheel steering and worked to get it right."

Kris pointed to the canard. "You also used the canard. You pulled back on the stick and that relieved pressure on the nose wheel that made the steering easier." I rotated the canard up and down. "If I did, I didn't do it consciously."

"On the next run concentrate on using the canard and the nose wheel. But do not rotate the stick back all the way. That will take pressure off the nose wheel and give you more control."

Down the runway again. Slight left pedal for the crosswind, half back stick rotation and a tapping toe on the right pedal to keep it on the line. We sped down the runway, nose wheel right on the centerline. Kris spoke over the phones.

"When I release at the end of the run, use the nose wheel and swing the nose across the runway."

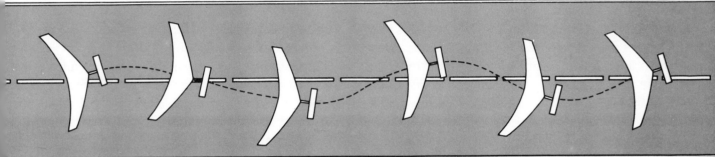

**MAKING PRECISE TAXI TURNS** back and forth across the runway while under tow requires positive nose wheel control.

Kris released and as we started to come to a stop, I rammed my left foot down hard on the pedal. The XL turned slowly and came to a stop with her nose pointing about 45° across the runway. Kris brought the nose around.

"This time we are going to try something different."

"Line yourself up on the dash marks on the center of the runway. Position yourself one foot on one side of the white line, then, while taxiing, cross over the line."

"That cross wind will tend to push the Eagle to the left. You'll have to fight the crosswind to maintain proper crossovers.

"On each crossover, straighten out, then cross over the line again. Repeat this maneuver until you get it perfect.

"Make the crossovers where you just barely go across the white line. Always keep the canard pitch down on these maneuvers. This will hold the nose wheel tight to the runway, permitting proper steering techniques. Stay close to the center of the runway.

"After each roll, slow yourself down by using the brake. Do not be afraid to drift off the runway when making a turn at the end of your roll.

"Watch out for overcontrolling on these practice ground rolls. Keep yourself lined up on the runway, not too far to the right or left. This is the trickiest part of the training, controlling that nose wheel. Keep centered on the line, then move one foot to either side of the center line."

We started down the runway. Gingerly I applied pressure to the right foot pedal. The Eagle nosed right. Again, softly, I pressed the left foot toe down as we gained speed. The bird zipped across the centerline so fast it caught me off guard. Instinctively, my right foot went down. Zing across. Left foot. Swish. We were much too far left. Right foot down to compensate. I was flying back and forth across the centerline in wide swinging arcs. I started to sweat as I desperately alternately applied left then right pedal to attempt to get the Eagle under control. When Kris released we were almost to the far right side of the runway.

I braked to a stop while in a half turn. The adrenaline was pumping into my blood stream. Kris walked up.

"What did you do wrong, Rick?"

I thought for a moment. "I overcontrolled. That is what I did."

"Right. To keep the aircraft stable you must be soft on the nose wheel control. Smooth it out on the next run."

The tow rope tightened as I shook the tension out of my muscles. Down the runway we went. "Don't overcontrol now, Rick," I thought. Very slowly I applied pressure on the right pedal, and as the Eagle started across the center line I again slowly pressed on the left pedal. She swung smoothly across the line as I started a series of right and left foot pressures. It was like dancing with a beat. As long as I kept the tempo all was smooth; then the crosswind caught me off guard. We swept far to the right.

We were heading for the grass on the side of the runway. If I didn't do something and quick, we were going into the turf. Left pedal and we swung back to the center of the strip but we went too far left. Bang. I hit the right pedal and relieved the pressure as soon as the bird responded. We were crossing the line when the tow rope went limp. Breaking to a stop, I wiped the perspiration from my brow.

"Whew," I thought. This is tough, but I felt good. Slowly I began to understand the timing and pressures needed to make the XL respond correctly.

"Like to take a break, Rick?"

I looked up at Kris.

"Not right now. I want to get that nose wheel under control before we break. Let's do a few more runs before we stop."

We made several more passes and by the time we finished, I felt that the nose wheel problem was licked. Little did I know.

We un-hooked the Eagle and I taxied her back to the hangar area with only a little trouble with the steering. We moved the Eagle into the hangar and went to lunch. It was 12:30 p.m. and I was very hungry.

☆

# PILOT BRIEFING

## AIRPORT-AIRPARK RULES AND RIGHT-OF-WAY PROCEDURES

Airports and Ultralight Airparks have right-of-way rules and procedures that are created to maintain a consistently smooth flow of air traffic into and out of the facility. Ultralight pilots must be aware of these rules and regulations so that he or she will become a *safety conscious* and well informed flyer.

Each airport or airpark will have a slightly different set of rules and regulations. These variances are included because of the prevailing wind direction, ground obstructions such as trees, power transmission towers, poles, buildings, mountains or hills, residential zones, local laws, lakes, streams, rivers and a host of other ruling factors that must be observed to maintain a safely run airport or airpark facility.

All airport and airpark facilities are built around a runway or runways. The main axis of most runways are usually laid out so that they point towards the prevailing winds. However, this is not always the case. Sometimes topography, obstructions or other controlling factors such as local laws or land boundaries determine the runway's direction.

Runways can be surfaced with concrete, asphalt or tarmac, gravel, crushed stone, dirt, grass or sometimes a plowed field. Each of these runway surfaces require pilot skill to land on safely. Practice and training can solve these specific landing and take-off problems.

All well-run airparks or airports have a "Wind Direction/Arrow Indicator," or "Wind Sock." In addition, ultralight airparks will have a recognition target mark located in a conspicuous spot.

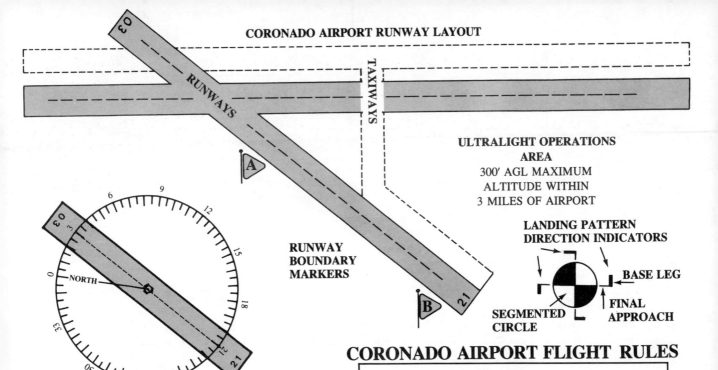

## RUNWAY NUMBERS

A LARGE WHITE NUMBER is painted on the approach end of every major airport runway so they can be easily seen and read from the air.

Runway Numbers are determined by the runways magnetic compass direction. For clarity, runway numbers are rounded off to the nearest tenth of a compass degree, with the deletion of the last zero of the compass heading direction. North on the Aeronautical sectional chart compass rose heads "0."

At Coronado Airport, the ultralight runway angles Northeast by Southwest. The numbers painted on the North East end read 21. On the South West end is the number 03. If we add the zeroes those numbers would read 210° North East and 030° South West compass headings.

Using only two numbers instead of three makes communicating between pilots and controllers easier to say, remember and identify.

### MAGNETIC COMPASS DIRECTIONS

| | | |
|---|---|---|
| NORTH | (360°) | = 0 |
| SOUTH | (180°) | = 18 |
| EAST | (90°) | = 9 |
| WEST | (270°) | = 27 |

The last two digit numbers are the ones used on charts, wind directions and radio signals throughout the world.

## CORONADO AIRPORT FLIGHT RULES

1. All Flight Operations MUST be conducted Southeast, East or Northeast of Marker Flags.
2. Towing Operations to the Southwest will terminate at the "A" flag area.
3. Towing Operations to the Northeast may begin at the "B" flag area.
4. Company vehicles only are allowed in the ultralight operation area.
5. Flight Operations will be within a ceiling of 300 ft. AGL while in the airport traffic area. (Three miles from the airport perimeter).
6. Stop and check the runway in use prior to taxiing out.
7. ALL first time flyers must obtain an airport briefing prior to flight activity.
8. Ultralight operators will notify UNICOM (or the flight department in the absence of UNICOM) of all activities to take place in the ultralight operations area prior to such activity.
9. All ultralights operating on Coronado Airport will have a functioning altimeter on board the craft.
10. During ultralight flight activity a member of the ground crew MUST have a radio with transmission capability on 122.8.
11. All vehicles will park on dirt portion located North of taxiway for 21-03.
12. All ultralights when not using the activity runway will remain inside the holding line and if possible in such a way to allow other aircraft to taxi onto active runway.
13. All personnel, unless instructed differently by a flight instructor, will remain off active runway at all times.

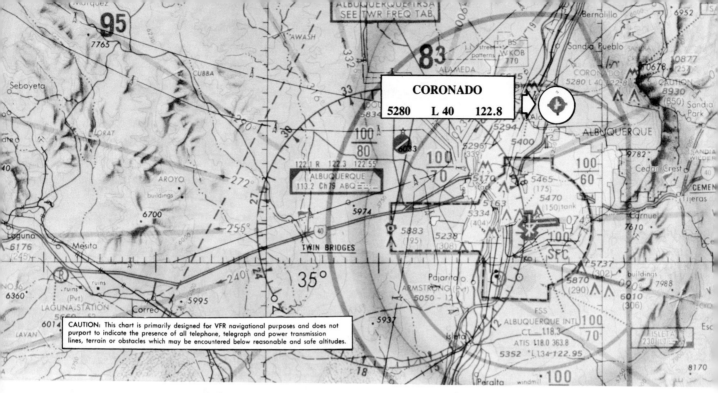

The airport shown on the sectional as a solid circle indicates that it is not tower controlled. The number 5280 is the altitude MSL. the letter (L) means the runway lights are on from sunset to sunrise. The star above the circular symbol with four nibs protruding around its perimeter designates that there is a rotating beacon that is lit from sunset to sunrise. The nibs tell you that service is available during daylight hours. The number 40 means the runway is 4,000 ft. in length and the number 122.8 is the UNICOM (Aeronautical Advisory Radio Station or FSS – Flight Service Station to call for weather and other information regarding airport services and landing conditions.

DETAIL FROM THE ALBUQUERQUE SECTIONAL AERONAUTICAL CHART

The Federal Aviation Administration (FAA), has divided all the airspace over the United States into Controlled and Uncontrolled sections. Specific sections are legal while others are illegal for overflights by general aviation which includes ultralights. These sections are clearly marked on Sectional Aeronautical Charts published by NOAA – National Oceanic and Atmospheric Administration and can be purchased for $2.75 ea.

UNCONTROLLED AIRSPACE – It's legal to fly within 700 ft. to 1,200 ft. AGL. There must be a minimum of 1 mile visibility and no flights into clouds.

VFR – Visual Flight Rules
Pilot must be able to see a minimum of three miles and the cloud ceiling must be at least 1,000 ft. AGL (Above Ground Level).

CONTROLLED AIRSPACE-TRANSITIONAL AREAS are located where the 700 and 1,200 ft. AGL zones intersect on the sectional charts.

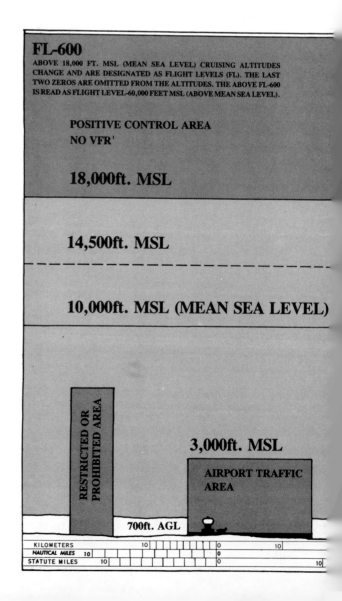

CONTROLLED AIR SPACE – It is not absolutely necessary to have permission, either by prior request or by radio contact with ATC (Air Traffic Controller), to fly in controlled air space. The "Controlled" means that the weather and visibility must be minimum VFR (visual flight rules), and you must look out for, and avoid, other aircraft in the vicinity of Controlled Air Space.

IFR – Instrument Flight Rules – Visibility is less than three miles and the cloud ceiling is lower than 1,000 ft. AGL. For ultralights it is illegal to fly in IFR weather. FAA Regulations 103.

To fly in Controlled Air Space in IFR weather conditions you must be certified within the last 6 months by the FAA as an IFR rated pilot, have an airplane that is equipped with IFR instruments, a 360 channel aircraft radio, and have filed an IFR flight plan with your nearest ATC (Air Traffic Controller).

FEDERAL AIRWAYS – Routes pilots on IFR fly during IFR weather conditions using VOR (VHF OMNIDIRECTIONAL RANGE – sometimes called OMNI for short).

CONTROLLED ZONES – Off limits to ultralights unless you have permission to enter them from the local ATC (Air Traffic Controller). The CZ extends from 0 (zero) AGL upwards to 3,000 ft. AGL.

TERMINAL CONTROL AREA (TCA) – Off limits to ultralights. Airlines fly in this zone.

AIRPORT TRAFFIC AREAS (ATA) – You must be able to talk to the ATC in ATA areas to get clearances to land or enter an ATA area. ATA has a 10 mile diameter from the center of the airport and extend upwards to 3,000 ft. AGL.

---

**CONTROLLED AIR SPACE – UNCONTROLLED AIR SPACE**

**VFR FLIGHT CONDITIONS** must be observed by ultralight pilots at all times while in the air. This includes strict observance of minimum distances from clouds, specific air space corridors, minimum visibility requirements from flight altitudes to the earth.

---

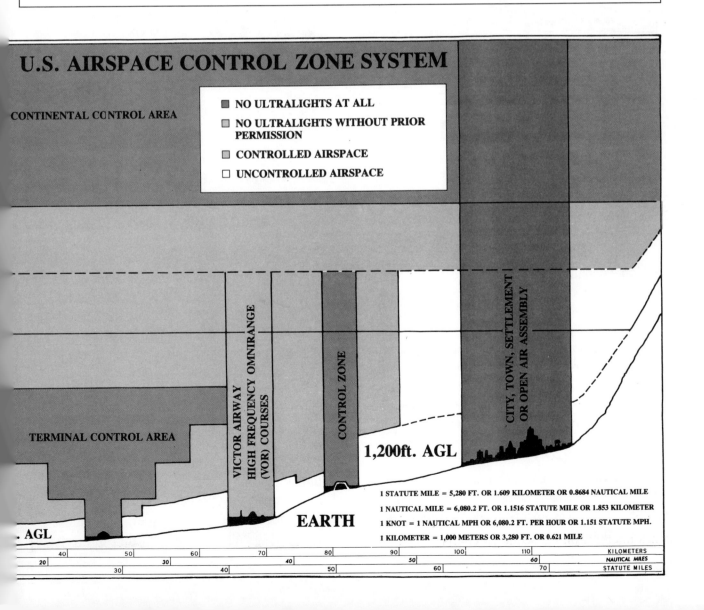

# ULTRALIGHT VISUAL FLIGHT REFERENCE (VFR) REGULATIONS

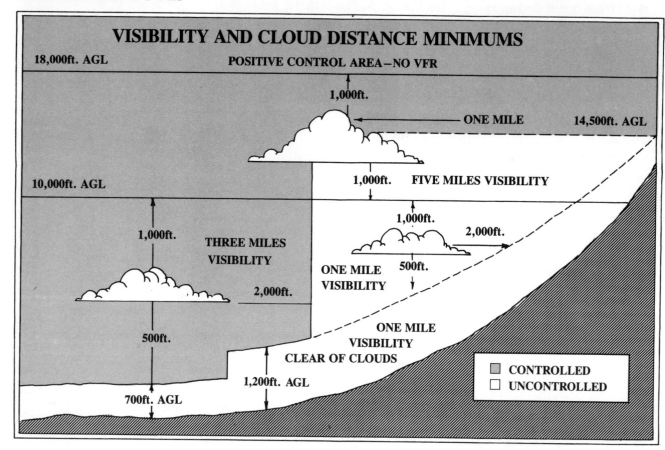

# AIRPORT TRAFFIC PATTERNS

## AIR TRAFFIC PATTERNS

All Air Traffic Patterns are designed around the rectangle of the runway. All incoming and outgoing air traffic directions are determined by the prevailing wind, and the air traffic pattern will change direction when the wind changes direction. From the air, the pilot can tell the air traffic direction by observing the wind direction either by the wind sock, ground arrow, smoke, waves on the water if nearby, treetops or field grass bending in the wind, ground dust or anything that can be observed that will indicate wind direction. Once the pilot knows the wind direction the landing procedure is then standard and is the same throughout the country.

## WIND DIRECTION

All takeoffs and landings are INTO THE WIND. Only in extreme emergencies is this rule broken. Taking off and landing with a TAIL WIND requires great skill and piloting.

## AIR TRAFFIC LEGS

The rectangular Air Traffic Pattern that bounds the runway is separated into 5 separate Legs.

1. UPWIND LEG—This leg parallels the runway and always points directly into the wind. Takeoffs and landings are into the UPWIND LEG. On takeoff, straight ahead maximum altitudes of at least 100 feet AGL should be attained before turns are made. Many accidents happen when pilots attempt turns before enough air speed and altitude are achieved.

2. CROSSWIND LEG—This leg cuts across the runway at a 90° angle. During landing approach procedures, many times the glide angle into the airports traffic pattern leads you "IN" on the crosswind leg. Altitude at this point is about 500 feet AGL or higher. When entering the CROSSWIND LEG scan the entire area for air traffic in the pattern or for those aircraft about to enter. The aircraft in front or below

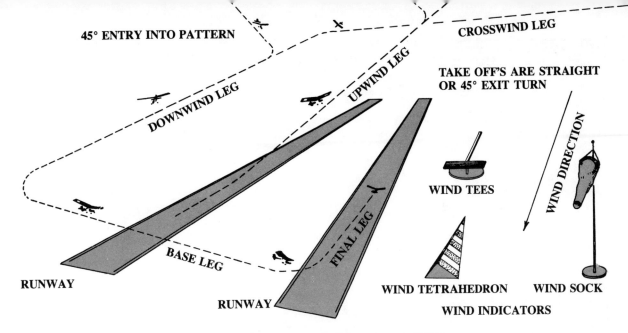

**AIRCRAFT ON FINAL LEG HAS THE RIGHT-OF-WAY**

you has the right-of-way (FAA REGS) so do not crowd or try to get in front. Be polite. Be right. If you must, go around and re-enter the pattern. Under no circumstances other than a *life and death distress emergency* should you cut in front of, underneath or around any aircraft in the pattern in front of you.

3. DOWNWIND LEG—Running parallel to the runway, the Downwind Leg is flown in the direction the wind is blowing which is opposite to the landing direction. It is commonly known as a Tail Wind. Most air traffic patterns are entered on the Downwind Leg at an angle of about 45° and at an altitude of approximately 500 to 300 feet AGL. At this altitude the pilot has an opportunity to stabilize the landing approach and correct for crosswinds. Also in the Downwind Leg, the pilot bleeds off airspeed and sets throttle adjustments for a gradual descent. It is important to stay well behind other aircraft in the pattern. Those airplanes in front have the right-of-way.

When you have picked out your touchdown spot on the runway and it is approximately 45° behind your shoulder while you are in the Downwind Leg, you then make the turn into the BASE LEG.

4. BASE LEG—Cutting across the landing Pattern at a 90° angle, the Base Leg is the critical landing approach decision point for the pilot. Once turned into the Base Leg, the throttle is backed off, flaps or spoilers are deployed to reduce speed and a commit to land is reached as the Final Approach Pattern turn nears. The Final Approach Pattern should be clear of all aircraft in the air or on the ground when in the Base Leg. Should an ultralight or similar airplane suddenly appear out of nowhere and fly into the Final Approach Pattern and you are floating in on the Base Leg, you must immediately get the throttle up and "pour on the coal," make a controlled pitch up and get away to the opposite direction of the intruding ultralight or airplane. Gain altitude and go around and line up in pattern to re-enter the crosswind or downwind leg at a 45° angle.

Gnash your teeth at the pilot who cut you off but remember this: You are in command of your aircraft, and your life. Take the smart way. Go around and keep a sharp eye out for the "Hot Dogs" or other runway jumpers. *Once you are safely down* is the time to tell off the hot jockeys, not in the air.

Barring all intrusions when in the Base Leg, you pick your landing spot and turn into the Final Approach Leg when the end of the runway is about 45° angle to your approach.

6. FINAL APPROACH LEG—This is the critical leg. So much so that the FAA regs stipulate that any aircraft on the Final Approach Leg has the right-of-way over all other aircraft in the air or on the ground. A pilot on "FINAL" is concentrating all of his attention on the landing and cannot be distracted in any way. This is why the FAA wrote the regulation. However, the safety-conscious pilot is always ready for any emergency should someone or something enter his final approach. Be ready to react and fly out and around. Do not be right according to the regs but dead because you insist on going on in the landing pattern.

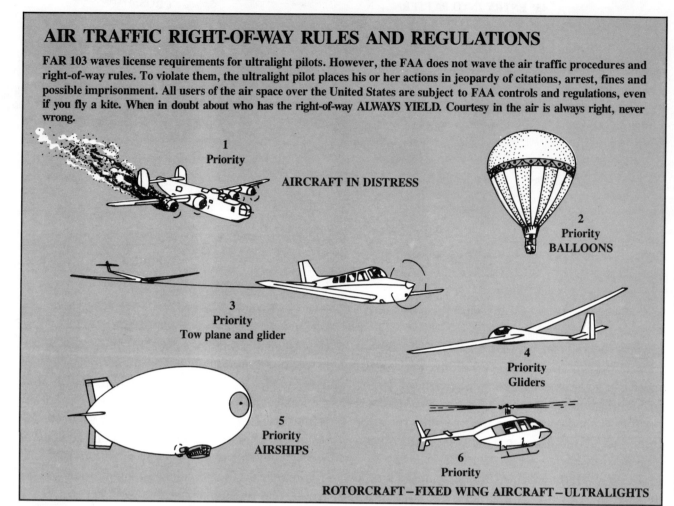

# AIR TRAFFIC RIGHT-OF-WAY RULES AND REGULATIONS

FAR 103 waves license requirements for ultralight pilots. However, the FAA does not wave the air traffic procedures and right-of-way rules. To violate them, the ultralight pilot places his or her actions in jeopardy of citations, arrest, fines and possible imprisonment. All users of the air space over the United States are subject to FAA controls and regulations, even if you fly a kite. When in doubt about who has the right-of-way ALWAYS YIELD. Courtesy in the air is always right, never wrong.

1 Priority — AIRCRAFT IN DISTRESS
2 Priority — BALLOONS
3 Priority — Tow plane and glider
4 Priority — Gliders
5 Priority — AIRSHIPS
6 Priority — ROTORCRAFT – FIXED WING AIRCRAFT – ULTRALIGHTS

## HEAD-ON APPROACH

When you are approaching other ultralight or similar aircraft in the air "HEAD ON", each pilot must turn right and pass on each others left.

Ultralights fly VFR (Visual Flight Reference) and many pilots fly along highways, railroads, power lines, river beds or towards clear landmarks. To be on the safe side of navigating these routes always use the same rules you would if you were driving a car on a highway, river or railroad. STAY TO THE RIGHT HAND LANE OR SIDE. If you fly down the middle you may run into another ultralight or other small plane doing the same thing. So stay to the right to be safe.

## RIGHT ANGLE APPROACH

When converging at the same altitude, the aircraft flying into your flight path from the RIGHT has the right-of-way. The aircraft on the LEFT must yield the right-of-way and turn to the RIGHT and allow the other aircraft to pass. DO NOT dive under climb over or pass in front of the approaching aircraft. The other pilot may try the same maneuver and you both can wind up in a collision.

## OVERTAKING FROM BEHIND

When overtaking another aircraft from behind, pass well clear of the slow aircraft on the RIGHT. The aircraft ahead has the right-of-way.

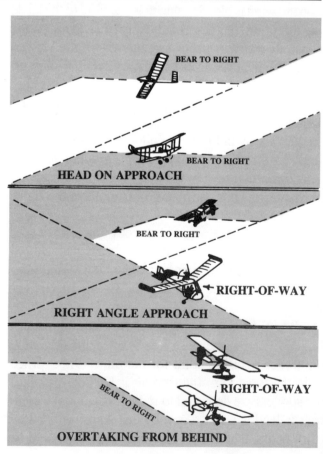

# FLYING HAZARDS

TALL TOWERS AND BUILDINGS — All tall obstructions present flying hazards. Crashing into them is the obvious problem for ultralight pilots but there is a more insidious danger surrounding tall objects; air turbulence downwind of the obstruction. Fly too close to one of these tall towers or buildings and you risk being twisted into a pretzel by the severe vortices created by the turbulence that always surrounds these hazards.

WIRES — Power cables, transmission tower support cables and the electrical transmission wires are the most critical of all hazards because from a distance they cannot be seen. A pilot flying at 45 to 55 mph will have at most one to two seconds of reaction time to take evasive action once the wire is seen. Flying under the wires is also a very hazardous maneuver. Accident reports tell the story. Many pilots who fly under wires get caught on the down wires that support the pole or tower. They cleared the horizontal wires but got hung up on the support cables.

The safest way to avoid wires is altitude. Hedge hopping can be fun but it increases your chance of tangling up in wires by 100%. Be wise and stay high. Leave the hedge hoppers to the birds, not ultralights.

RIVERS — Waterways can be a hazard when the ultralight pilot flys too low over the water. Crossing rivers are bridges, pipelines and power lines. Again, the maps do not show these obstructions except the bridges, so beware of flying low over the water. It is fun. It is exciting. It is dangerous. It is deadly. It is a stupid thing to do. Don't fly low unless you are landing.

HILLS, VALLEYS AND MOUNTAINS — Wind turbulence around hills, valleys and mountains can be extremely hazardous when encountered by an ultralight. A thorough understanding of how winds and turbulence are created is a must for the ultralight pilot. Study Meteorology and Micrometeorology and spend time observing your flight zone before blindly venturing into unknown territory.

STORMS — Any storm, foul weather, gusty or high winds should ground an ultralight.

BIRDS — Many airplanes have been destroyed when encountering birds in flight. A bird flying into a propeller or into the pilot can have tragic results. Stay well clear of roosting sites and bird flocks. Most birds will clear off a noisy ultralight but watch out for the stray or curious ones.

DOMESTIC ANIMALS — Loose dogs, horses and livestock can be a severe hazard to an ultralight. They have no knowledge of what a whirling propeller can do and when an engine is fired up, the noise can frighten animals and cause them to dash or run right into the ultralight.

HUMANS — People are curious and will crowd around an ultralight and not know how dangerous the power train and propeller of an ultralight are. Keep people well away and always shout loudly "CLEAR PROP" before starting up.

AIR PARKS AND AIRPORTS — Taxiways, hangars and vehicles can be the cause of mishaps. Taxiing into another aircraft is not an uncommon occurrence, either on the taxiway or in the hangar. Ultralights that have not been "TIED DOWN" have been demolished by high winds. Runway and Taxi surfaces with chuck holes, rocks, gravel or debris can damage the ultralight. Stear clear or land on an unobstructed part of the runway or strip.

GET-THERE-ITIS — The pilot becomes the hazard when good judgement and common sense are overruled by an urgency to fly somewhere when the flight conditions dictate "stay put" and the pilot overrules and takes off, many many times with fatal results.

# FLIGHT TRAINING – EAGLE XL  0700 hrs

**DAY 2**

The exhilarating experience of the first day's flight training was still churning in my mind as I began the morning ritual of mentally preflighting the Eagle XL. Chris had the bird out of the hangar and was in the process of attaching the left wingtip rudder when Bob and I drove up. As we got out of the car Bob spoke.

"Today you will experience flying the Eagle one to three feet in the air, under tow, of course."

He opened the door of the training center and asked if I would like a cup of coffee before we began the day's lessons. Thanking him, I declined and walked up to Kris; I was roaring to go.

"Rick, you can begin your preflight."

He stepped towards the training center doorway.

"I'm going inside. Call me when you think the plane is ready to fly." Kris was all business. I liked him and his technique. I lowered myself and started examining the nose wheel and all of the attachments there. I really enjoyed the preflight. It gave me a chance to carefully examine and touch the entire airplane, a liberty that is not allowed when you are looking at other airplanes. Pilots and owners are very fussy about strangers messing about with their airplanes. Touching is a definite no-no, unless it is your own airplane or have permission from the owner or someone designated to look after the airplane.

**KING POST TENSIONER NUT**

I'd just finished examining the teleflex cable and the next item was the King Post attach point. I raised the King Post Tensioner Oversleeve. The King Post Tensioner Nut was half way down the threaded end of the King Post. I tightened it back up against the tensioner and slid the Oversleeve back down. I wondered if Kris had planted any more traps for me. By the time I had completed the Preflight, I had found everything to be in order. Kris walked out of the hangar as I spoke.

"The King Post Tensioner Nut was not in the right position. I tightened it up." I slid the Oversleeve up and pointed to the nut. Kris felt it, making certain the nut was properly seated. It was. He slid the Oversleeve down as I asked.

"The Eagle is ready to fly, but to be certain, did I miss anything?" I asked.

Kris smiled. "No. Warm up the engine before you taxi out to the flight line."

I set the choke, yelled out "Clear Prop!" and pulled the starter handle. The Cuyuna barked as it turned over. I reset the choke and let her warm up before running up the engine. Kris lay my helmet on the seat of the Eagle and held the cockpit cage as I ran up the engine to full throttle, held it there for a moment, then brought the throttle back down to idle, then hit the kill switch. Kris and Bob helped me straighten out the shoulder harness straps as I settled into the seat, buckled up and put on the helmet.

Kris leaned forward as he spoke. "Yesterday, Rick, you did fine. Today we will continue with the slow and fast tows and work on nosewheel control. After you learn nosewheel control we will get you off the ground. Not high. Only a few inches at first. I'll be using the radio so when I get on the tailgate be ready to taxi behind us."

I nodded my head as I fastened the helmet chin strap. Kris's voice came over the headset.

"Rotate the canard if you are receiving me." I did. "Good. Now follow us a short distance behind and keep the nosewheel on the center line."

I wiggled the canard, reached behind, it was a hard awkward stretch but I reached the choke, depressed it, grasped the starter pull handle, yelled "Clear Prop!" and yanked down hard. The Cuyuna caught immediately. Again I stretched hard back and returned the choke to normal setting.

Starting the engine before the taxi set the juices flowing. I looked out across the wings and felt a feeling of excitement ripple through my body. The vibrations from the engine made the fabric covering the wings dance. I held my foot on the brake pedal and ran up the throttle. The Eagle strained against the foot brake. I released and started the taxi.

Sitting at the controls of an airplane rolling across

the ground under its own power is a powerful experience. It is unlike any that you can have while driving a vehicle. An airplane can fly. A car, bicycle, motorcycle, boat cannot fly; and that is the difference.

When you taxi an airplane you are driving it towards the active runway where you can increase the throttle and thrust and lift into the air. That is what an airplane is. A flying machine. And you know it the moment you strap yourself tight into the seat.

When you run up to an active runway and cast your eyes skyward, looking for airplanes landing and taking off and you have never flown before, makes you think seriously about the mission you have committed yourself to. You are on your way into the sky and that reality sets serious thoughts into your physical actions. You become very cautious, alert, sensitive to changes and other sensations never before experienced.

*An airplane is only as good as its pilot.* That thought kept slipping into my mind as I worried the nose wheel of the Eagle XL along the yellow taxi line. Be gentle and easy was the thought that constantly slid into my thinking as we swung around the ninety degree curve. I moved off the yellow line only three feet instead of fourteen, as I did yesterday. I was beginning to understand the driving foot pressures needed to steer the Eagle XL.

My eye lifted from the slick surface of the front wheel tire and fixed on the taxiway. I focused tightly on Kris sitting on the tail gate of the stationwagon. I looked at the turf around them. Port and starboard. I liked using nautical terms. It made me see clearly when I used them.

Clumps of scrub grass spotted the entire area around the airport. There were no trees higher than five feet unless they were grouped around buildings.

The sky above—High Cirrus with a jet contrail streaking west towards San Diego—Slight ground fog. Surface winds by the slant of the windsock: one-to-two mph. South westerly.

I checked my hands. The left one was on the throttle. I tightened my fingers around the throttle handle and pushed my arm forward. The Cuyuna rattled behind me. I shook my shoulders loose and settled my feet on the rudder pedals. I had to immediately get the nose wheel under control. We thrust forward with power.

Loud mechanical sounds roar all around you penetrating the helmet padding and earphones. They are close noises. The exhaust blasts of the gasoline engine is only a foot or so from your ears. Your body vibrates with the Eagle as you bounce along the gravel. Everything shakes. Good reason for that careful preflight. One safety pin off and the item it was holding would surely shake loose, either here on the ground or worse, in the air. Prime experiences for a fledgling pilot, I thought.

I cleared my head, looked at the horizon up ahead and kept that nosewheel on the yellow line.

Taxiing the Eagle XL is sheer pleasure. I listen to the Cuyuna engine chugging along nicely as we bounded along the taxiway. I glanced at the wingtips dancing up and down, reflecting every slight bump or depression in the taxiway and they reminded me of the Rockettes.

I concentrate on the nosewheel. I am in charge of its direction. I depress the left and right rudder pedals and notice the time it takes to wiggle the nose back and forth. I concentrate and adjust the throttle to maintain speed as we crunch along the yellow line.

**NOTE HOW THE TOW CABLE** is covered with a plastic tube to prevent any abrasive action during the tow training exercises.

**THE TOW-CABLE IS ATTACHED** to each side of the flight cage at junction points that are able to withstand the stresses applied during flights under tow.

**THE MOMENTS OF WAITING FOR THE TOW CABLE** to snap taught and drag you down the runway are moments to gather thoughts together about the specifics of aircraft control. For me, the thoughts were about the steerability of the nose wheel.

Before I knew it we were on the flight line. I was hooked up. The tow line tightened and snapped and I was hurtling at a steep angle toward the runway.

I straightened out to a heading straight down the middle of the runway. The spotted surface of the tarmac flowed under my heels. My concentration hooked up to my feet and the nose wheel steering. I flexed my feet hard and quickly against the rudder pedals and began to feel the control of the Eagle as she accelerated quickly. I missed the rattling noise of the Cuyuna while under tow.

The speed increases. We start to squirrel in quick oscillations port and starboard. I work the rudder pedals to soften the pressure and get her back under control.

I can feel myself tightening up at the speed. I glance at the tow line and then the car. Kris speaks. "Keep it straight, Rick." Straight down the runway. The speed increases. The wings shudder and I tighten muscles. Wheels rustle. White runway dashes flash underneath in flickers.

You begin to become quite concerned. You are strapped into an airplane, hooked to a car with a rope that is accelerating down a runway and you aren't quite sure what to do next. Kris speaks.

"Rotate slowly back on the stick just a little and rotate the canard down." I did. The nose wheel lifted and floated in the wind. My eyes were glued to it. I couldn't take my eyes off it. The Eagle lifted into the air. All three wheels were off the ground. I instinctively pushed the stick forward and we settled back on the surface of the runway.

The rope slackened. I let the Eagle roll out to her natural stop. I kept my feet still on the rudder pedals. I did not attempt to steer the bird. She came to rest on the edge of the runway. Her nose wheel was setting on an anthill. The ants were swirling around my nose wheel.

This was almost like being at an amusement park. We lined up on the runway. Car speeds up. You fly up. Float over the runway. Feel the breeze in your face. Know you are in the air. Eyes scan the wingtips. They are flying. You are flying. You are only one inch off the earth.

You do it again and again and it feels the same every time. Great. I want to do it again.

Shoulders back tight against the seat. Hold the stick forward. Keep the nose on the deck. Hard worries about foot control on the rudder pedals. We are moving very fast. Twenty-seven MPH. Little mistakes lead to bad moments. So take it easy even though your teeth are glued together.

Kris speaks. "Level the wings to the horizon. Keep them there. That's good. Now set her down as we release the tow."

Gawd. The terror of the lash up. The tightening of the cable. It shatters the dream. Flying was not the reality of a cable snapping tight and hauling my body rapidly down a runway. Flying was a float. Not a snap.

I was learning. Flying was a lot of things that have nothing to do with flying.

Kris spoke. "Hold it about a foot off the runway and keep it centered over the runway dashes."

I'm up and over the dashes. Flip wrists port, starboard, center to keep her wings level. Really subtle corrections made with quick decisions. But right ones made instinctively which made me feel I might become a flyer.

Sitting on the end of the runway. A Thought. *To fly or not to fly. That is the question.* What do I think of before that cable snaps tight and I have to start making fast decisions.

Relax. Take it easy. Everyone says that. But I don't feel that way.

I grit my teeth and sweat it out as the cable tightens and yanks me down the runway.

I'm flying. Everything worked out. I'm in the air. What an experience. It's like a friendly hug. It is better than I thought.

I'm beginning to think of my umbilical cable as a friend, even when it is slithering out ahead of me and I have to get it together on the cable's schedule, not mine. Instructions are hard to follow.

Flying's like that.

When the cable snaps tight. The moment is there. Tires roll. Nosewheel straight. Pick up speed. Rotate the stick back.

You are in the air. You know you are only up there for a few seconds at most. The freedom is spectacular. Flying is spiritual.

**TIP RUDDERS CONTROL YAW.** Note how the Eagle is yawing to port and how the port tip rudder is angled inboard to effect the yaw maneuver.

**FIRST FLIGHT: LOOK CLOSELY** and you will see the wheels just inches off the earth. Note how I have the tip rudders deflected. This rudder control was not intentional. I was trying to shove my feet into the nose wheel, toes first. Chris soon corrected that bad habit of mine.

You float in the air only inches above the runway. Rotate back very slowly, wrist fluttering to keep the wing level. Rivet the eyes on the end of the runway.

Pull back on the stick. Mush. Touchdown. When the Eagle stops, I unbuckle and relieve myself on the open desert.

Airborne. Flying four inches AGL (above ground level). Pull back. Up she goes. Push forward. Down she goes. Level off. She smiles at me. We are flying. 4" AGL.

This is it. I've made it.

Survived.

Just pay attention to Kris the instructor and all is OK.

Coming in for a landing. Speed right. Nosewheel cockeyed. Touchdown imminent. Correction. Port? Starboard? Decisions. Which way?

Pilot decision. Level the wings. Nosewheel straight ahead. It takes a little right rudder deflection to get it on the line. Take a breath and set her down. The nose wheel straight ahead, spinning madly to a stop.

Kris speaks. "We're going to do it again. Keep working on maintaining level wings like you've been doing, and keep your eyes on the horizon not on the runway. If you don't watch the horizon you can lose lateral control and not know it."

"I must remember to do that," I thought. "I have been looking at the nosewheel too much."

In the air again. Flying. Floating over the ground. Air flowing across the eyes. Eyes scan the horizon, wingtips, altitude. Time flys, so am I. My feelings at the exact moment of flight, another quick scan of everything and down. Let out the air. We're down and safe.

That time we flew 5" AGL. The thrill is always there every time you lift off and touch down.

I'm six inches AGL. I rotate the control stick forward. The nosewheel kisses the tarmac. It is turned to starboard. The Eagle dashes hard towards the edge. Kris releases the tow line. The Eagle keeps right on speeding towards and off the runway and down into the ditch to a stop. The nose was dug into the roots of a scrub bush. My heart was pounding. Kris rushed up.

I unbuckled my helmet strap and slid it off as I unhooked and crawled out of the seat.

"You OK?" Kris looked closely at my eyes as I punched the palm of my hand with my fist.

"Damn," I said. "It just kept on going off the runway and I froze tight. I just let her run into the scrub and I don't know why."

Bob came up and looked the Eagle over carefully.

**THE TAXI ROLL-OUT** at the end of a tow run is the time to get your breath back. These first sessions of tow training did put my nerves on edge. But my instructor sensed this and constantly talked the tension out of my body.

He looked at me and smiled. "Looks like there was no harm done. Let's push her back up on the runway."

Kris and Bob wisely let me calm down as we eased the Eagle back up onto the runway. I set about pre-flighting the bird. By the time I was through I was calm. I walked up to Bob and Kris standing beside the station wagon. I said what I was thinking.

"My mind was like a computer that went on overload. Suddenly I simply froze. Nothing worked. I let her run off the runway without doing a thing. And I think I know why." Kris spoke. "Don't worry about it. It is part of the training experience." "Kris, what happened was that I was having too much fun and my mind was off the business of flying. I was enjoying the experience and when the plane started off the runway I was not prepared to react properly. To say it simply, I was not thinking. My mind was somewhere beside flying the plane."

"I'm glad you said 'Flying the Plane.' That is the key word and you said it."

"Kris, my right foot simply kept on pressing down. I'd lost all sense of which was left and which was right. It was a weird experience to say the least."

"Well you've learned something today about flying. First, always fly the airplane and second, keep your mind on the business of flying."

I interrupted. "Right. Keep your mind on the business of flying when you are flying."

I picked up the helmet and put it on and sat down in the seat of the Eagle and buckled in.

"I've had my break and like they say 'When you fall off the horse, get right back on and ride it again.'"

Kris said: "OK. But what we are going to do now is go back to the slow and then the fast tows. Keep the canard slightly back so the entire pressure of the plane is not on the nose. That will keep the plane from 'Squirreling back and forth.'"

We went up and down the runway while I perfected the steering technique and relearned which foot was left and which was right.

I thought about the run-off for quite a while before I got to sleep that night. But I had flown. Only three feet off the ground but I was up there. It was with that thought that I drifted off into a deep sleep.

I was ready when Bob beeped the horn at 0630 hrs. and my third day of flight training began.

0700 hrs
Weather: Overcast
Wind: Light South Westerlies

# FLIGHT TRAINING—EAGLE XL

**DAY 3** Bob Mullikin delivered a quiet student to his instructor. Kris Williams had the Eagle XL out of the hangar waiting for my preflight. I stroked the wing of the Eagle and I followed Bob to the coffee urn that was tucked neatly beside Brian Allen's office. Bob spoke as he poured the coffee Kris had thoughtfully prepared earlier.

"Today, Rick, we will do the low and medium tows. That is about three to six feet in height. Also you will practice flares and "S" turns. This is the hard part of your flying lessons. These low and medium tows are the most difficult and demanding. They require concentration and precision controls, which by the way, you have."

His words reverberated in my mind. I knew I must pay close attention to instructions.

Outside, Bob looked up at the overcast sky. "Very little wind today but it looks like we may have some rain." Kris spoke.

"I don't think it will rain right away. Perhaps later this afternoon but we can get in a few tows before then."

During my preflight I found a safety pin missing from the left wingtip rudder pivot arm and that was all. I pointed this out to Kris and he handed me the missing pin. I replaced it, warmed the engine and had a good taxi to the flight line.

My hot enthusiasm was quiet now. I was thinking too much about what I was doing to get excited. Now was all business. I had the steering problem under control. I tracked that nose wheel right on the yellow taxi line all the way to the active runway crossing. Only once did I veer off the yellow line. It was only a few inches when negotiating the ninety degree turn. I was beginning to get the feel of the Eagle XL's personality.

Try as I might, I had trouble keeping my adrenalin down when the Eagle accelerated down the runway behind the tow car. Instinctively I tightened up. I knew this was not good and vowed to get that muscle tightening under control. Everything must be soft and easy. No jerky moves. That was the instruction and that was what I was going to achieve. It started to drizzle.

Kris asked. "Does training in the drizzle bother you?" My answer was positive. "The rain does not bother me one bit. I can go as long as you think it safe to continue. You're the boss."

Kris looked up the runway and to the sky as he said, "You must be careful when stopping on the wet runway. The foot brake requires a lot of pressure because the tire is wet and has no tread. It is smooth."

It was just a fine mist that streamed across my cheeks as we reached flying speed. 26 mph. Kris spoke to me.

"Now rotate back on the stick and raise the nose. That's good. Now hold it there and notice when the mains break away from the runway. Good. Now keep the wings level and rise up to about three feet off the runway. Good. Raise the right wingtip a little; Use left spoileron. Whoops. You used too much and you held it there. Use just a little correction. You used too much. You are getting too high. Bring it back down to three feet. Use slight forward stick pressure. Good. Now keep it centered over the white centerline. You are too far left. Use the right rudder. That's good, Rick. You used just enough. Keep it centered. You are beginning to raise up and drift off center."

I began to flutter my wrist back and forth, raising and lowering the port and starboard spoilerons and keeping my eyes on the end of the runway. We swung over the white line to the right. I flicked my wrist to left and oscillated the stick in short quick moves back and forth, all the while keeping my eyes glued to the end of the runway and peripherally on the wing and the horizon. I wanted to keep the wings level and yaw the plane back to the center of the runway and those wrist movements did it as Kris released the tow.

The main wheels touched first as I followed Kris's instructions to pull back gently on the stick as the Eagle settled down towards the runway. We lost airspeed and the nose wheel bumped down and my concentration went directly to steering the nose wheel, all 100% of it.

The roll out after landing is exciting. You still are in command of an airplane. She bumps and switches as you come to a stop, where you want to stop, and not somewhere else.

I was where I wanted to be when Kris and Bob came up to me.

"You did fine, Rick. You're getting the feel of the controls. I watched you working the spoilerons up and down to keep the wings level and on course. The landing was fine. Now let us do it again and this time swing back and forth over the center line. Go just about three feet on either side of the line and start the crossover when you are about six or so feet above ground level. Now a little breeze is picking up. Notice the wind sock and compensate for it when you are airborne.

"Now, I am not going to tell you when to rotate back on the stick. You sense when you have enough airspeed to lift off and make the proper move."

**TOW CABLE** tightens and we accelerate.

**26 MPH** and up in the air we go—4 feet AGL. Starboard spoileron deflection leveled the wing.

**STARBOARD YAW**—Then port yaw to bring the Eagle over the centerline for the touchdown.

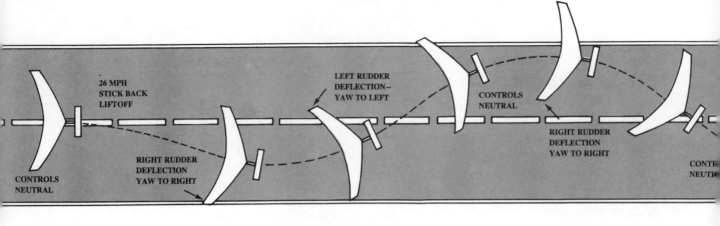

**"S" TURNS—TRAINING MANEUVERS WHILE UNDER TOW.** These exercises require exact coordination of both feet and hand on the stick. In the beginning there is a lot of over control. With practice, coordinated turns are mastered.

The next few flights were progressively higher and wider on the "S" swings back and forth over the runway. I was beginning to notice how the perspective of the runway and earth changed as you pitched up and down and yawed from side to side. Also I took special note of the time lag from control signal to airplane response. This took concentration. There was no place for idle thoughts here. Every move had to be anticipated before action so no quick jerky moves could occur. I learned this by getting into serious trouble.

The drizzle was about the same. Just a little mist. It didn't bother me. The takeoff roll was right. Liftoff at 26 mph. Wing angle of attack set perfect. We sail up to about ten feet. We're flying straight down the center of the runway. I glance at the tow car. Chris, peering up at me, sat very still. The angle of the tow rope seems extremely sharp. My altitude is now about 14 feet AGL. I glance down at the runway and hold my attention on the white dashes falling behind. I hold the attention too long. In the last few seconds that my concentration is on the runway markers my port wing started to tilt down into a roll and kept on going until the turning force of the Eagle snapped my attention away from the runway.

I looked up. We were in a 10° bank to the left and falling fast towards the ground. Kris released the tow rope immediately. Just as I started corrective moves, I remembered being told to always fly the airplane and to maintain airspeed. My body reacted.

The port wing tip rudder was first to graze the ground. I rammed hard right on the spoileron to bring the right wing tip down and brought the stick back into a flare set. A moment later we hit. The Eagle was just starting her corrective moves when we went in. We bounced hard into the sandy shoulder of the

**ROPE TIGHTENS.** Down the runway you go.

**FLYING SPEED IS REACHED.** Rotate the stick back and . . .

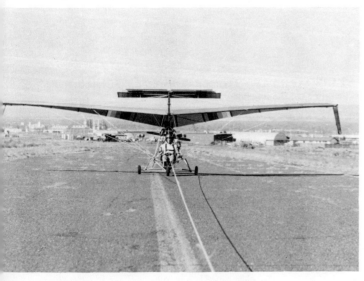

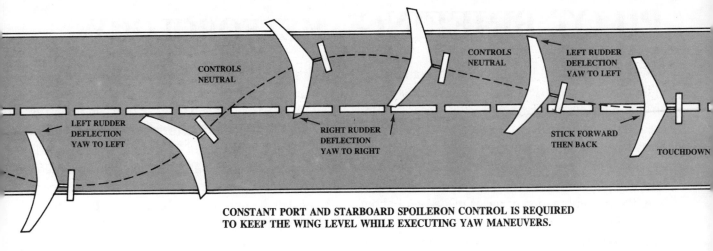

**CONSTANT PORT AND STARBOARD SPOILERON CONTROL IS REQUIRED TO KEEP THE WING LEVEL WHILE EXECUTING YAW MANEUVERS.**

runway and slewed to port, pivoting around the dragging wingtip rudder, and came to rest with the nose wheel dug into the sand.

Bob and Kris were quickly beside me. They helped me unbuckle and get out.

We looked at the Eagle as Kris asked me if I was OK. I said I was and we carefully examined the Eagle. We all moved her back up on the runway and started to study the left wing tube. Kris seemed to think it was slightly bent. Bob did too so Kris preflighted the Eagle and took her up for a check flight. He went around the runway doing Touch and Goes, then landed and taxied up, killed the engine and got out.

It seems to fly straight and true. Not out of trim at all. I guess that port leading edge wing tube didn't get bent or if it did it didn't change the trim.

We discussed the incident in detail and concluded that my not paying attention to flying the plane especially on lift-off was the mistake. Fortunately, not a serious one – this time. But never let it happen again. I also did the right thing in keeping the nose down and levelling the wings before the flare move.

I got back on the seat and we did five more runs up and down the runway and stopped only because it started to rain.

During the taxi back I was deep in thought and steering the Eagle was tricky with its wet nose wheel. I felt that on this day I had had a deep learning experience. One that I knew would keep for the rest of my life.

**UP YOU GO.** Total concentration is mandatory. Otherwise...

**YOU FIND YOURSELF IN THE ROUGH.** In this case undamaged.

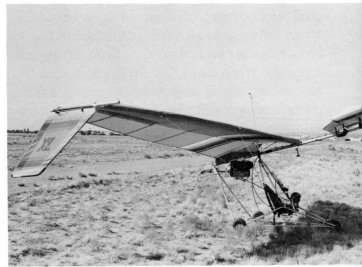

# PILOT BRIEFING – METEOROLOGY

We will begin this briefing with an examination of the odorless, colorless, tasteless liquid gas called air that blankets the Earth with a thickness of hundreds of miles. It is this atmospheric system that supports flight and in combination with heat produced by solar light, creates the weather, life and death on Earth.

AIR – A combination of gases made up of about 20.9% Oxygen ($O_2$), 79.1% Nitrogen ($N_2$) and a small amount of Carbon Dioxide 0.033% ($CO_2$). Plants absorb the carbon dioxide for growth and release oxygen and a minute quantity of other rare gases as a byproduct of the interaction.

These atmospheric gases are stabilized in a constant percentage in the Earth's entire boundary layer of atmosphere, from MSL (Mean Sea Level) to the limits of the Geocorona.

An invasive molecule is also present in the atmosphere in varying percentages: $H_2O$, Hydrogen Oxide (Water). An unstable molecule, it is lighter than $O_2$ or $N_2$. As $H_2O$ is heated by Solar light it changes into a gas and rises. At high altitudes it is cooled and the molecule returns to a liquid (water) and falls to earth as rain, snow, hale or sleet.

When an air molecule, in still air, makes contact with the moving airfoil molecule, the air molecules are captured and held in place by the magnetic field of the molecules that make up the composition of the airfoil. Aluminum, dacron, wood, paint, fabric, or whatever the airfoil surface is made of, are all composed of molecules.

Once the air molecule is parked in orbit next to an airfoil molecule, it remains there unless it is dislodged by compression and heat. When a high enough temperature is reached, the oxygen molecule ignites and the molecule holding the oxygen molecule in its magnetic field burns along with the oxygen molecule. When the oxygen fire is completed, the oxygen and airfoil molecule have changed their chemical composition. They become free of magnetic attraction and are blown away in the wake turbulence of the moving airfoil.

Examples of this molecular interactive destruction were demonstrated when the space capsule called "Skylab" incinerated upon re-entry into the atmosphere. Its descent to Earth left a trail of ionized gas and debris of burnt-up matter scattered across Australia's surface. The recent space shuttle has a protective shield of ceramic tiles to protect it from a similar destructive end upon its frequent re-entrys into the Earth's atmosphere.

The Boundary Layer of Earth Atmosphere extends from mean sea level (MSL) vertically to an altitude of 620 miles where a zone called the Exosphere establishes its outerlimits. Beyond the Exosphere, the atmosphere thins down into the Earth's Geocorona that measures several radii of the Earth deep.

The weight of the atmosphere is less than a one-millionth part of the Earth's mass. A one square inch column of our atmosphere, extending from the Earth's surface (MSL – Mean Sea Level), upwards into the space continuum, weighs only 14.7 (PSI – pounds per square inch).

Three fourths of the total mass of the atmosphere is located below 35,000 ft. (6.63 miles) above (MSL – Mean Sea Level), the altitude commercial jet liners regularly fly. Fifty percent of the atmosphere is below 18,000 ft. MSL (3.41 miles), where general aviation regularly flys. At 1,000 ft. MSL where ultralights regularly fly we have about forty-nine percent (49%) of the total mass of the Earth's atmosphere.

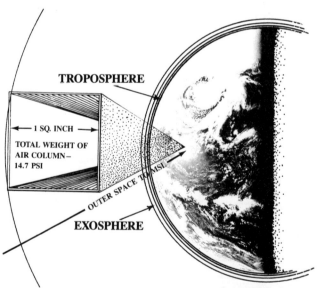

The Earth's boundary layer of atmosphere is composed of a series of spherical laminal envelopes containing layers of gases, vapors and suspended matter, bound to the Earth by gravitational force (G).

**TROPOSPHERE** - Extends from the Earth's surface to an Equatorial altitude of about 8 miles (13 kilometers, 42,240 ft.)
5 miles (8 kilometers, 26,400 ft.) Polar, and is the limit of the weather. Cirrus clouds form in this region.

**EXOSPHERE** – Extends from the Earth's surface to an Equatorial altitude of 620 miles 10,000 kilometers.

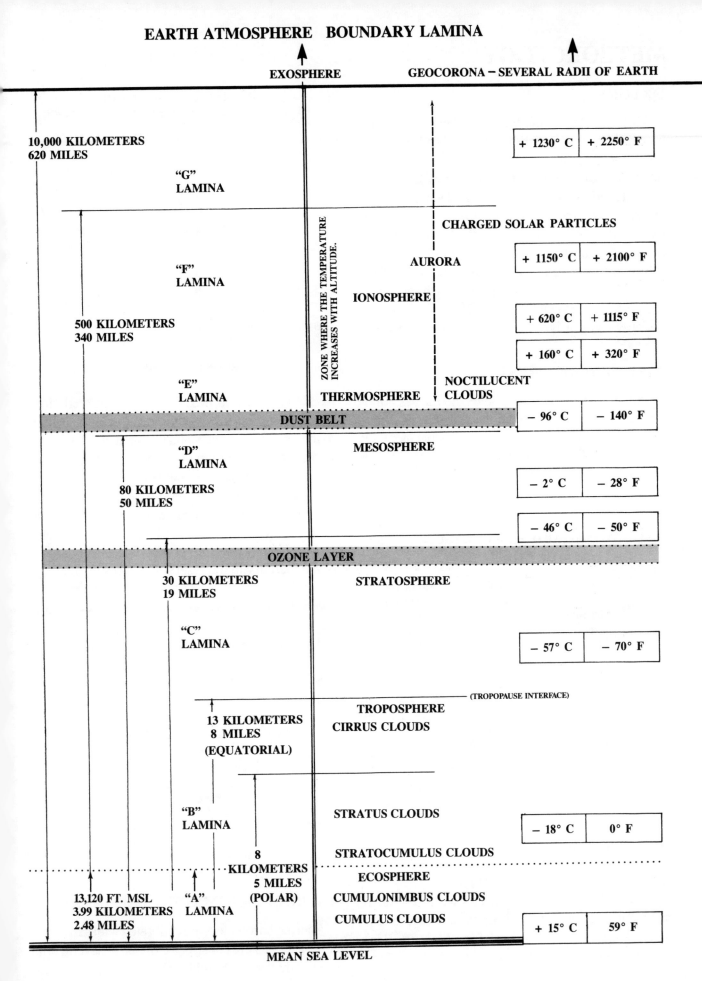

# METEOROLOGY

## OXYGEN

All life forces on the Earth are dependent on the boundary layer of air for their survival and existence. Of all the gases that make up the air, oxygen is the only one that supports life as we know it. It is the inner reaction of this oxygen molecule to other molecules that makes life work. Oxygen is produced by photosynthesis generated by the leaves of plants, evaporation of water (It is estimated that there are 4 billion tons of oxygen in one cubic mile of sea water.) and by chemical interaction with the elements.

By weight oxygen 23.15%  20.98% volume.

The weight of igneous rocks is 46.6% oxygen.

## ATMOSPHERIC PRESSURE

The force of gravity (G), holds the Earth's atmosphere to our planet and, in the process, also produces atmospheric weight and pressure that increases and decreases with changes in temperature, humidity and air mass circulation.

### STANDARD AIR PRESSURE AND ALTITUDE

| Altitude (Feet) | Volume Pressure (Inches Mercury) | Parts per Pressure (Pounds per Sq. Inch) |
|---|---|---|
| 50,000 | 3.44 | 1.7 |
| 30,000 | 8.88 | 4.4 |
| 25,000 | 11.10 | 5.4 |
| 20,000 | 13.75 | 6.8 |
| 15,000 | 16.88 | 8.3 |
| 10,000 | 20.58 | 10.1 |
| 5,000 | 24.89 | 12.2 |
| 1,000 | 28.86 | 14.2 |
| Mean Sea Level MSL | 29.92 | 14.7 |

BAROMETER (BAROS, Greek for weight) — An instrument that measures atmospheric pressure (P). There are two types of barometers; Mercury (Hg) and Aneroid.

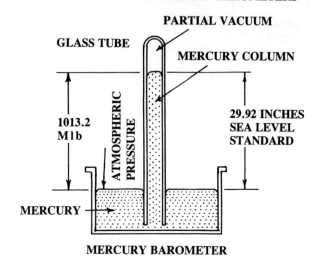

**OPERATING FUNCTION OF THE MERCURY AND ANEROID BAROMETERS**

MERCURY BAROMETER

MERCURY BAROMETER — A calibrated, long slim (3 ft.) glass tube with one end sealed, is filled to the top with mercury and the open end is inserted into a reservoir of mercury. The weight of the mercury in the tube is drawn down by gravity (G) and flows into the reservoir until it equalizes the force of gravity by a partial vacuum which the descending mercury created in the sealed top of the glass tube.

Atmospheric pressure (P — 14.7psi) on the surface of the mercury in the reservoir, forces the mercury being supported by the partial vacuum in the sealed end of the tube to move up and down in precise phases with atmospheric pressure variables.

The mercury will rise or descend in a pressure lapse rate of one inch (34 millibars) per 1,000 ft. from MSL. 1,000 millibars = 1 bar = atmosphere = 14.7 PSI.

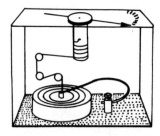

**ANEROID BAROMETER**
Invented in 1643 by Torricelli.

## ANEROID BAROMETER — BAROGRAPH

Aneroid barometers are calibrated with a mercury barometer and are constructed of one or several thin, usually metalic collapsible capsules called aneroids.

The capsules are hollow, partial vacuum sealed discs with bellows surfaces. They are spring loaded to support and balance the external pressure of 14.7 PSI and the partial vacuum inside the aneroid capsules. The capsules expand or contract in phase with atmospheric pressure changes.

Mechanical linkages connect the aneroid capsules to a read-out scale. When the atmospheric pressure falls, the aneroid disc expands and a meter needle moves to the left. When the pressure increases, the aneroid capsule compresses and the mechanical linkage moves the needle on the read-out scale to the right.

**ANEROID BARAGRAPH**—A record of time and pressure is recorded on a slowly rotating drum. This is the type used to record ultralight altitude records. Note the NAA security seal.

Every pilot must be able to interpret barometric readings in conjunction with additional meteorological data when reading weather reports and in determining flight plans and operations.

Pilot Dick Rowley (EEA #148288) set a world altitude record in his kit-built Mitchell-U2 Superwing ultralight. A barograph was installed, and sealed in the nose of his ultralight. It recorded, on paper, the exact altitude of the U2 attained and later confirmed by the National Aeronautic Association (NAA) and the Federal Aeronautic International (FAI). The Barograph trace proved the U2 reached an altitude of 25,940 ft. above MSL. The temperature at that altitude was −30°F. on September 17, 1983. Location—Colorado.

Airplanes float in the air as do ships at sea. Airplanes and ships have many things in common. They both have: captains, pilots, crews, navigation, logs, hulls, decks, bulkheads, a port, starboard, inboard, outboard sides, rudders, propellers, sails and many more.

The Earth's atmosphere is in many ways similar to water that covers almost 71% of the earth's surface.

Water and air are both liquids, and they have in common: temperature, turbulence, wakes, storms, waves, cold and warm fronts, streams, curls, oxygen ($H_2O$) and many more.

## BEAUFORT WIND SCALE

| Code Number | Wind Velocity (mph) | Description |
|---|---|---|
| 0 | 0-1 | calm |
| 1 | 1-3 | light air |
| 2 | 4-7 | light breeze |
| 3 | 8-12 | gentle breeze |
| 4 | 13-18 | moderate breeze |
| 5 | 19-24 | fresh breeze |
| 6 | 25-31 | strong breeze |
| 7 | 32-38 | moderate gale |
| 8 | 39-46 | fresh gale |
| 9 | 47-54 | strong gale |
| 10 | 55-63 | whole gale |
| 11 | 64-75 | storm |
| 12 | over 75 | hurricane |

Developed by Admiral Beaufort of the British Navy to determine the effects of wind speed on canvas sails of ships. It is still in use to identify and classify wind speeds and force by number.

# METEOROLOGY

In addition, the Earth's water in the form of surface liquids and atmospheric gasses, are heated when radiated by solar light. Solar light radiation contributes about 1% of the total atmospheric heat. The balance of atmospheric heat is generated where the air molecules occlude or make contact with solar heated Earth surfaces.

The atmosphere descending from cool high altitudes by the force of gravity (G), slowly gains density as it compresses. (Page 84)

When the air molecules of the descending air touch the solar heated surface of the Earth, they expand, increase in: temperature, humidity, loose density and weight and become buoyant. The buoyant atmosphere rises into the sky vertically by a push-up force exerted by dense cool air. A balloon soaring into the sky is the result of this force: dense air forcing thin air up and away.

In a stable atmospheric environment this up and down cycle of air movement is contained within layers. (Page 83)

In an unstable atmospheric environment, these rising and descending columns of air break through the atmospheric layers of air and cause unstable weather. (Page 89)

Solar light when deflected into the atmosphere from the Earth's surface and rotation produces all our weather.

High atmospheric and Earth surface temperatures (T) are produced in the zone where the solar light is at or near a vertical angle of 90° over the Equator.

Absorbed and radiated atmospheric heat diminishes as the solar light incident angle increases from the 90° vertical at the Equator to 180° at the poles. (Page 87)

Solar light and Earth conductive heat vaporizes humidity from the air and Earth surfaces. The vaporized water is transformed into ice crystals at high altitudes, clouds, hail, rain, snow, sleet and drizzle.

WEATHER—Atmospheric conditions influence human behavior, such as: Hot or cold, rain or snow, dry or humid, wind or calm, fog or fair or a hurricane or a breeze. All these weather conditions are of vital importance to the pilot.

Human survival above the Ecosphere, 13,120 ft/MSL = 2½ miles, is not possible without supplemental oxygen being added to the flight support system. At this altitude, atmospheric oxygen pressure is not enough to prevent headaches, fatigue, faulty judgment and lack of coordination. Rapid progressive deterioration of the human system prevails if deprived of oxygen at high altitudes.

*FAR 91.32 excerpts: Above 14,000 feet MSL the pilot must use oxygen continuously for any flight exceeding 30 minutes, during the day. At night, the pilot must use oxygen at altitudes above 5,000 feet MSL.*

Solar heated surface air at the Equator loses humidity and density as its temperature increases. The heated molecules of air rise into the atmosphere into a vertical Equatorial Trough.

The pressure of the rising air column in the Equatorial Trough, forces the circulating cooler air at the top of the trough to split into two large Northern and Southern hemispheric circulating tropical air masses. These tropical air masses are then drawn down the Earth's curvature, by gravity and cool air at lower altitudes, towards the cold, dense, atmosphere at the North and South poles. The frontal edges of these tropical air masses are called Tropical Fronts.

On their way to the cold poles, the Tropical Front rapidly lose heat, humidity (as rain or snow) and altitude. The atmosphere gets cold and heavy as the density increases.

Reaching maximum weight and density at the poles, the atmosphere slowly builds up into a large polar dome that is drawn back up over the surface of the Earth's curvature towards the warm tropics.

The frontal edge of this Polar air mass is called a Polar Front.

The clean, cold, dry and heavy Northern and Southern Hemispheric Polar Fronts continually snake their way up the slope of the Earth's curvature towards the lighter warm updrafts of the tropics in an endless back and forth cycle. Two other forces influence their journeys: Friction from the Earth's surface and the Coriolis Force, also known as the Geostrophic force or planetary rotation.

Surface friction from the Earth's seas and land account for a counterclockwise Northern hemispheric atmospheric veering deflection of 5° to 10° for the seas and 10° to 15° for the land. In the Southern hemisphere these same atmospheric percentages are "Backed Off" in a clockwise direction.

At the poles, 90° N-S, The Earth's rotational speed is slight but as we move down in latitudes to 0° at the Equator, the diameter is 8,000 miles and the rotational speed has increased to 1,000 (600knt) MPH.

The spinning Earth's Coriolis or Geostrophic Force effect on the air molecules occluding with the Earth's surfaces drag at and deflect the air molecules 90° at a right angle, counter-clockwise to the wind gradient blowing from high (Equatorial) to low (Polar) air masses and back.

The Coriolis force is also responsible for the counterclockwise swirl of water draining out of a sink or tub (in the Northern hemisphere) as it descends. In the Southern hemisphere the reverse happens.

The Earth's surface friction force on the cold, heavy, dry, low pressure moving Polar Front, extends vertically 1.5 to 2,000 feet AGL before reaching the friction free air stream. Wind velocities in this friction zone (33 ft. to 2,000 ft. above the Earth's surface) are influenced by the air stream type; Dry, clean, cold, unstable Polar Front and wet, dirty, hot, stable Tropical Front air masses.

The combined effects of the Earth's friction and the Coriolis Force on the Polar Front moving towards the Equator separates it into six individual circulating atmospheric cells; three in each hemisphere.

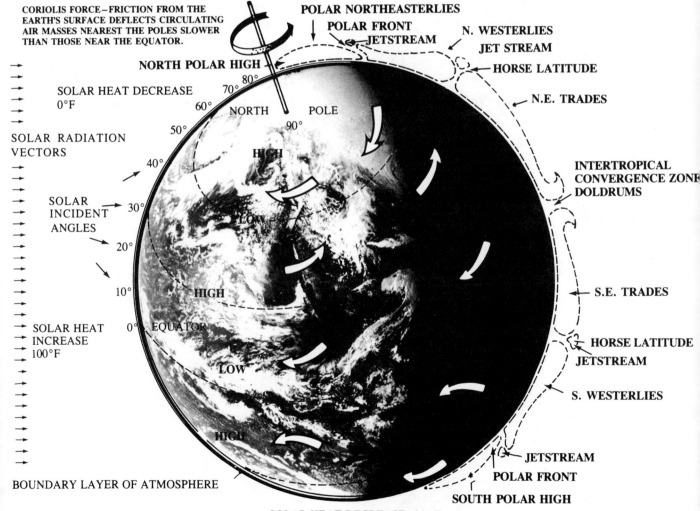

EARTH'S SURFACE – 196,938.800. sq. miles
EARTH'S SEA AREA – 145,000,000. sq. miles
EARTH'S LAND AREA – 51,938.800. sq. miles.

# METEOROLOGY

The six atmospheric cells the Coriolis Force creates are called:

POLAR NORTHEASTERLIES—Circulate clockwise from the point of origin (90°N to 60°N Latitude) and blow inbound (to the West).

NORTH POLAR FRONT—Separates the Prevailing Southwesterlies and the Polar Northeasterlies (40°-60° North Latitude) in a trough of Eastward-flowing cyclonic swirls of rain clouds and variable weather.

PREVAILING SOUTHWESTERLIES—Circulate counterclockwise from the point of origin (60°N to 30° North Latitude) and blow outbound (to the East).

HORSE LATITUDES—Separates the Prevailing Northwesterlies and the Northeast Trades at (30°-40° North Latitude) in a trough of calm, light variable winds. Sailors called these the Horse Latitudes when sailing ships were becalmed and had to throw their livestock into the sea to conserve fresh water for their own survival.

NORTHEAST TRADES—Circulate clockwise from the point of origin 30° North to plus or minus 0° Equatorial.

INTERTROPICAL FRONT CONVERGENCE ZONE (Also known as the DOLDRUMS)—Separates the Northeast Trades and the Southeast Trades in an Equatorial Trough of calms, variable and occasionally violent squally thunderstorms.

The SOUTHEASTERLY TRADES, HORSE LATITUDES, PREVAILING NORTHWESTERLIES and the POLAR SOUTHEASTERLIES flow in opposites as their counterparts to the North.

AIR MASS—There are three types of air masses; POLAR AIR MASSES are formed over the poles and move towards the warm tropics. MARITIME AIR MASSES are formed over the seas and move towards the Low Polar air masses. CONTINENTAL AIR MASSES are formed over the continents and move towards the low pressure air masses.

AIR MASS—Air masses have either low or high pressure fronts and are formed when noncirculating air builds up over the cold polar zones (low front), deserts (high front), forests (low fronts), continents (high fronts) and tropical oceans (high fronts).

The temperature and humidity of an air mass is determined by the seasons and environment of the region it forms over. An air mass can have a volume of thousands of square miles and be several miles thick. Its temperature is usually fairly constant from MSL to about 5 miles AGL.

As the air mass system moves across the Earth (about 350 nautical miles per day) it brings the weather it was formed over to the region along its trajectory.

AIR MASS MOVEMENT—When an air mass reaches its maximum size and starts to move, they are classified as either warm or cold by the

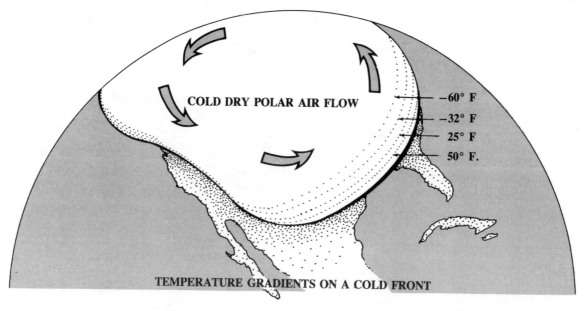

**TEMPERATURE GRADIENTS ON A COLD FRONT**

temperature of the Earth's surface they were formed in and move across.

A cold air mass moving down from the polar zone crosses over warmer Earth surfaces. Heat is transferred from the warm surface into the cold air mass and this action produces convective currents.

CONVECTIVE CURRENTS of hot air cool to the dew point as they reach altitudes of 8 to 10,000 ft. AGL and condensation occurs which form cumulus clouds, storms, rain, snow, hail and dust storms.

A WARM AIR MASS that is cooled by the Earth's surface keeps most of its temperature and humidity except for a few thousand feet AGL. Warm fronts usually have smooth, calm air, stable internal air currents and form long banks of low stratiform clouds after the warm front passes. In the summer the sun usually burns off these stratus clouds. In the winter they drift for miles before dissipating.

FRONTS — Cold and warm air masses do not tend to mix. When they do interface or come together, they form a Frontal Zone or what is commonly called a Front.

Fronts create extremes of weather. When a cold polar meets a warm tropical front they generate; storms, high winds, cyclones, hurricanes and general bad weather. Flying near or even miles away from these interfacial or transitional zonal front lines that can extend for thousands of miles can be extremely hazardous for the aviator who ventures too close to their violence.

A thorough understanding of these fronts is mandatory for anyone who is considering flying in the Earth's atmosphere.

COLD FRONT is a cold, dense air mass that hugs close to the Earth's surface as it advances. When a Cold Front interfaces with an approaching Warm Front, the cold, heavier, dense air pushes under and shoves the lighter, less dense, moist air of the warm frontal air mass upwards. This interaction is called a Cold Front.

In the Northern Hemisphere, cold frontal lines extend from the Northeast to the Southwest and move in an East, Southeasterly counterclockwise direction at about 25-35 MPH.

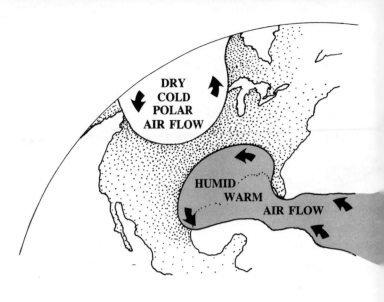

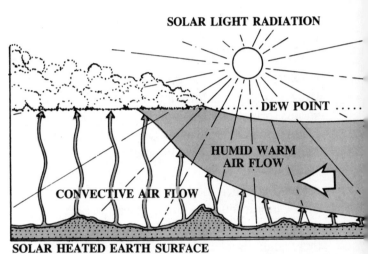

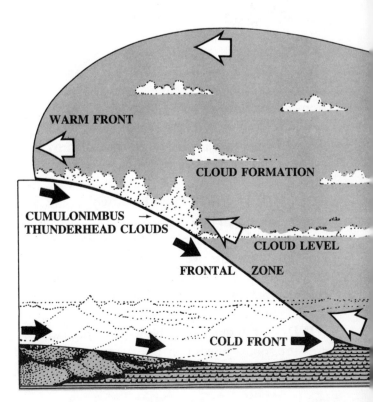

# METEOROLOGY

**DEW POINT**
**HUMID WARM AIR FLOW**
**COLD DRY POLAR AIR FLOW**
**COLD FRONT**

**WARM FRONTS** – When a warm air mass, moving at approximately 20 MPH, catches up to a cold air mass, the warm air slowly blankets up over the cold air, held close to the Earth's surface by weight and friction. This interaction produces a Warm Front.

When the humid, hot, Tropical Maritime Front rises vertically at the Polar Front, clouds of high frozen ice crystals form into wisps of Cirrus-Cirrostratus formations at altitudes of 20-40,000 ft. AGL. Cirrus clouds herald a low depression advancing at about 500-1,000 miles Westerly.

At times, Cirrus cloud formations can be misleading when they are composed of the remains of a Cumulo-numbus thunderhead breakup.

Contrails from high flying jet aircraft produce these contrails when the air is saturated with moisture and the water vapor from the jet engine condenses. Contrails can be an indication of Tropical air moving into the region in the absence of Cirrus cloud formations.

**STATIONARY FRONT** – When a cold and warm air mass meet and because they have equal pressure, stop, a Stationary Front is formed.

Winds along a Stationary Front blow parallel to the interfacial line and the weather created by this event is similar to a Warm Front.

**OCCLUDED FRONT** – When two Polar air masses, separated by a depression, force their way under the high front and meet, the two Polar Fronts violently push the warm tropical air upwards. Severe, unstable weather occurs all along the Occluded Front, an extremely hazardous flying zone.

**WIND** – Air in motion and blows from high to low pressure zones. Winds may be dry, moist, warm or cold and moves over the Earth's surface in swirls, drafts and changes in direction that at times can be violent.

In parts of the world the wind can blow in one direction for days or weeks, while in other parts the wind can shift direction and velocity continually.

Wind, constantly moving at variable speeds is influenced by: Atmospheric Pressure (P), Coriolis Force and surface friction, that deflects the winds to the East in the Northern Hemisphere and the reverse in the Southern. The interaction of these forces is called the Geostrophic Wind (turning world).

Winds are caused by: Frontal movements, surface friction and topology, Earth's rotation, differences in atmospheric temperature, air pressure and the continual interaction at the interface of the High and Low frontal zones.

Wind velocity increases with altitude. On a calm day on the surface of the Earth the winds may be two to three MPH. Aloft over the same point the winds can be 300 MPH.

**WIND DIRECTION** – Wind directions are classified by the compass point the wind blows from (32 directions) and is recorded to the closest 10° of the 0° to 360° compass scale. Example: A Northeasterly wind blowing from .09° NNE is expressed as a NE wind.

**TRUE WIND** – Compass point the wind is blowing from, regardless of the airplane's flight direction.

**RELATIVE WIND** – Wind direction the aircraft is always heading into, regardless of the True Wind direction.

**BACKING WIND** – Blows counterclockwise to the main, true wind direction. Example: NE to N.

**VEERING WIND** – A change in wind direction, clockwise to its direction. Example NE to SE.

ANABATIC WIND — Blows up the slope of hills and mountains as a result of solar heat on the Earth's surface. Begins ½ hour before sunrise and ends ½ hour after sunset, with a velocity of about 14 mph.

GRADIENT WIND — A horizontal wind that blows at a constant atmospheric pressure along the surface of the Pressure Gradient. It does not change its direction or energy but the wind speeds can vary.

GRADIENT PRESSURE and WIND are indicated on weather maps and charts with lines called ISOBARS.

ISOBARS — Equal Barometric pressure points are connected with lines that are spaced about four millibars apart. The Isobar stream lines, never the same on two maps, define the contours of the High or Low Pressure System.

Isobar lines are similar to the contour lines on a terrain map. They show the steepness of elevations and depressions. When the lines are close together, the gradient is steep. When they are far apart, the slope is gradual.

ISOBAR SYMBOLS — Arrow points for Low Fronts and half circles for High Fronts show the Gradient wind direction and are spaced along the frontal line either close together for high winds or far apart for Low winds.

Winds always flow parallel to the Isobar lines that show the boundaries of equal pressure along a High or Low pressure system.

## PRESSURE SYSTEMS

HIGH PRESSURE SYSTEM — Air from a high pressure system flows clockwise and is surrounded by a Low Pressure System.

LOW PRESSURE SYSTEM — Air from a Low Pressure System flows counterclockwise and is surrounded by a High Pressure System.

RIDGE — A knife edge ridge extending upwards of high pressure that runs from the center of a high pressure system.

TROUGH — A valley of low atmospheric pressure that runs from the center of a Low Pressure System.

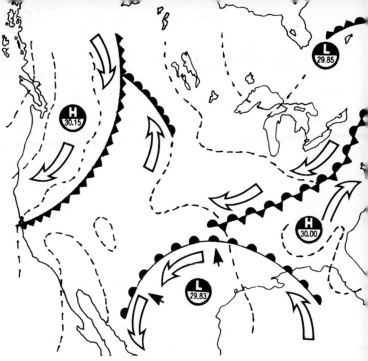

WEATHER MAP DETAIL SHOWING ISOBAR STREAM LINES AND WEATHER SYMBOLS

# WEATHER SYMBOLS

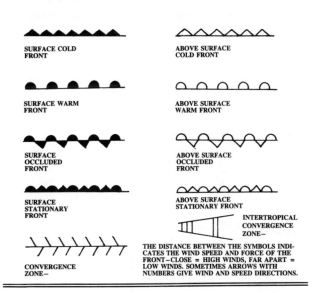

AIR DENSITY — Air, when subjected to high temperature, low atmospheric pressure and altitude becomes thin and less dense.

DENSITY ALTITUDE — Barometric pressure and temperature are used to compute *Density Altitude*.

As the density of the air *decreases*, Density Altitude *increases*, which lowers the performance of an aircraft's engine, propeller and the wings lifting capabilities. Longer takeoff and climb distances are required under *Density Altitude* flying conditions. An airport located at 1,000 ft. MSL may have a Density Altitude elevation of 7,000 ft. MSL.

# METEOROLOGY

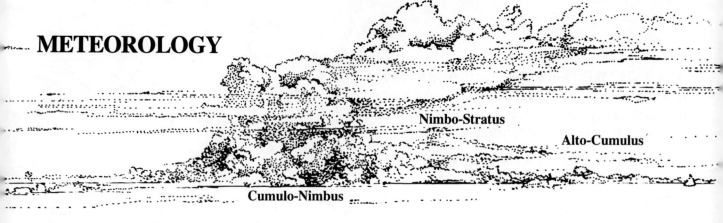

## CLOUD FORMATIONS

THERE ARE THREE basic cloud types out of a total of ten classifications and are named according to their (genera) altitude and cloud family.

| CIRRUS | STRATUS | CUMULUS |
|---|---|---|
| High | Medium | Low |

CLOUDS are made up of water droplets, ice crystals or condensation cooled to the Dew Point (100% moisture saturation) and are held in suspension in the atmosphere.

There are ten main cloud classifications and are named according to their (genera) altitude and family.

### CLOUD FORMATIONS

| Altitude—MSL | | Names-Genera | |
|---|---|---|---|
| 20-40,000 ft. | HIGH | Cirrus | Ci |
| | | Cirro-Stratus | Cs |
| | | Cirro-Cumulus | Cc |
| 6-20,000 ft. | MEDIUM | Alto-Stratus | As |
| | | Nimbo-Stratus | Ns |
| | | Alto-Cumulus | Ac |
| 0-2,000 | | Stratus | St |
| 1-5,000 | LOW | Cumulus | Cu |
| 1-4,500 | | Strato-Cumulus | Sc |
| 1-5,000 | | Cumulo-Nimbus | Cb |

It is essential for pilots to understand cloud formations and what they mean when planning flight operations.

THUNDERSTORMS—A thunderstorm can develop at any time of the year although they are usually associated with the Summer and fall months. They are the most devastating weather conditions for ultralights and general aviation combined. Thunderheads are formed after the sun has warmed the Earth and air at low levels and very low temperatures at high altitudes. The moisture in swiftly rising thermals of hot air condense into ice crystals and form Cumulus clouds. When a cold Polar Front occludes with an advancing Tropical Warm Front, severe thunderstorms and cyclonic winds are the usual results at the point where the two fronts collide. Once a developing thunderhead reaches 60,000 ft. AGL an anvil-shaped top of the cloud forms and drifts in the direction the thunderhead is moving. The anvil is formed of ice crystals.

Inside the thunderhead storm cell, intense up and down drafts traveling at 50-70 miles per hour fall and rise as much as ten or eleven miles. The down drafts strip moisture from the air and falls to Earth as rain. Some of the rain drops do not penetrate the base of the storm cell and are blown back up inside the cloud. Up and down they go and in their journey form into balls of ice called hail. When they are of a size and weight to resist the up and down drafts the hail blows out of the storm and falls to Earth. Hail stones have reached the size of grapefruit in some storms. Any aircraft flying into one of these hail storms can be wrecked in a few moments. Within an hour the storm blows itself out. The remaining clouds left over from the thunderstorm still have punch. Vortices of hundreds of mph churn throughout the atmosphere after the major portion of the thunderstorm has passed.

When thunderstorms are predicted, it is the wise pilot that waits it out on the ground, with the Ultralight securely tied down.

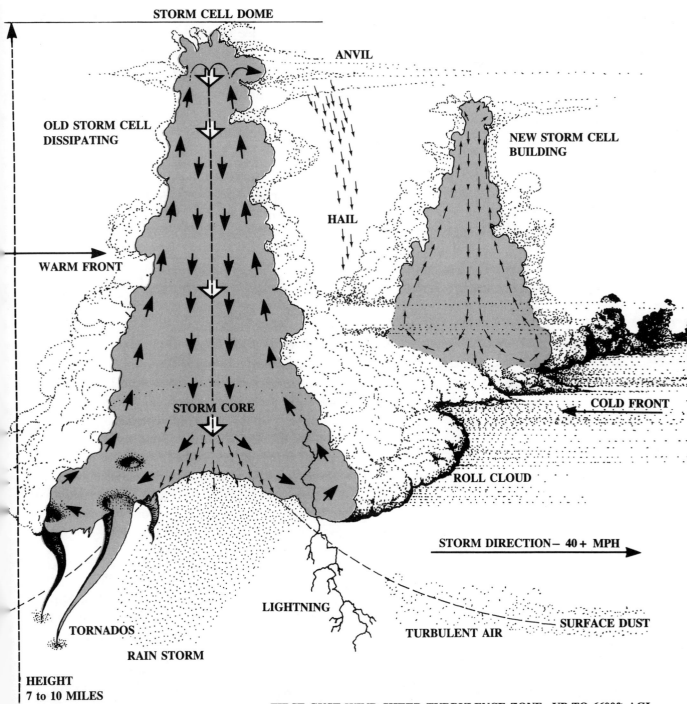

# METEOROLOGY

BREEZE – A light or moderate wind. Land breeze blows from land to sea when the sea water is warm and the land is cold. Sea or lake breeze blows from the sea or lake towards the land when the land is warmer in the day and the sea or lake is cold.

BUYS BALLOT'S LAW – Professor Buys Ballot in 1857 stated "Stand with your back to the wind; Low Pressure is on your Left and High Pressure is on your Right. In the southern hemisphere the reverse is true."

BLIZZARD – Strong winter snow storms with winds up to 75-120 M.P.H.

CALM – Absence of air in motion or when the air is blowing at less than 1 mph.

CHINOOK – Also known in locals as ASPER, NORTHWESTER, PUELCHE, SKY SWEEPER, ZONDA. A warm, dry wind on the lee side of mountains caused by the air becoming warm by compression as the air descends the mountains' encline slope after the air passes over the mountains' crest (not to be confused with a Katabatic Wind).

CONVECTIVE CURRENTS – Updrafts or downdrafts produced at the boundarys between land, water, deserts, forests, above cities, large buildings, hangars, mountains, along squall lines and thunderstorms.

| Convective Currents | Convective Currents | Squall Lines Thunder-storms | Obstructions |
|---|---|---|---|
| Gust Velocity – ft./per sec. | 1-35 | 35-130 | 5-50 |
| Altitude in Feet – AGL | 0-20,000 | 0-60,000 | varies by type of obstruction and prevailing wind. |

DOLDRUMS – Light variable winds at the Equator. At times severe thunderstorms can occur in the Doldrums also known as: INTERTROPICAL FRONT CONVERGENCE ZONE.

DUSTSTORM – A severe, strong dust-filled wind that covers a large area, usually found over deserts and can reach altitudes several thousands of feet thick. Also known as: DUSTER or BLACK BLIZZARD.

FALL WIND – A strong cold wind that blows down a mountainside and is the opposite of a foehn wind.

FOEHN WIND – A down slope, warm wind that blows on the lee side of a mountain slope.

FREE ATMOSPHERE – The atmosphere above the boundary layer of air where the Earth's surface friction is absent. The free air base is known as the GEOSTROPICAL WIND LEVEL.

FRICTION LAYER OF AIR – The air below the free atmosphere where the Earth's surface friction and rotation influence the flow of air. The friction layer of air is about 1,500-3,000 AGL.

GALE – Wind velocity ranging from 32-36 MPH.

GUST – Brief, transient, sharp increase in vertical or horizontal wind speed. An upward gust can lift an airplane or wing quickly as though going over a bump. A downward gust will lower the angle of attack and cause the airplane to sink as though it were falling into a pocket or hole. The density of the air remains the same in gusty conditions. Gusts are serious forces in ultralight flight.

Every ultralight pilot should practice emergency stability controls under manageable flight conditions so when a sudden gust catches you will be prepared for the emergency.

JET STREAM – A world-circling, rapidly moving narrow tube of air that blows in the stratosphere in an Easterly direction. Jet Streams circulate in the corridor where Tropical and Polar Fronts occlude (come together) and measure several miles wide at about a mile thick. Jet Streams travel at a velocity of about 50 to 400 knots per hour (58 to 520 M.P.H.) from 20 to 60 thousand feet AGL.

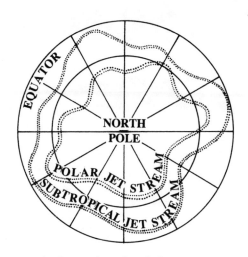

Jet streams in the northern hemisphere.

MECHANICAL TURBULENCE – Airflow over and around ground obstacles.

WIND SHADOW – The zone between two obstructions where the wind velocity decreases.

WIND SHEAR TURBULENCE – An instantaneous change in wind speed and direction in either a vertical or horizontal thrust that can violently alter an airplane's angle of attack and altitude. Wind shear turbulence can happen at any altitude but for the ultralight pilot, the low level wind shear is one of the most hazardous flying experiences a flyer can have.

Wind Shear Turbulence can invariably be found at the Interfacial zone between a rapidly advancing cold front (30knts or more) and a warm tropical air mass. The differences in temperature of 10° or more between the colliding fronts causes severe wind roters that blow in unpredictable directions.

There are four primary low level wind shear zones: Rapidly moving Fronts, temperature inversion zones, obstructions and thunderstorms.

When an airplane enters a wind shear zone it loses airspeed and lift. The plane sinks and unless there is enough altitude to regain air speed by advancing throttle an untimely collision with the ground occurs.

Extensive cloud formations develop along advancing fronts. The ultralight pilot must learn how and where these clouds are formed and be knowledgeable in recognizing the dangers with these fronts.

KATABATIC WIND – Blows down the slope of a mountain or hill. Also known as: GRAVITY WIND. Produced by surface cooling along the slope or incline which makes the air more dense. Begins ½ hour after sunset and ends ½ hour before sunrise, with a velocity of about 10 mph.

LOCAL WIND – Small area winds that differ from the movement of the main air mass pressure distribution.

MOUNTAIN AND VALLEY WINDS – Winds that blow along a valley's axis, then blow up the valley and hills during the day (14 +/− mph) (Anabatic Winds); starting ½ hour before sunrise and end ½ hour after sunset.

At night the winds begin (Katabatic Winds) blowing at (10 +/− mph) as the Earth cools, beginning ½ hour after sunset and ends ½ hour before sunrise.

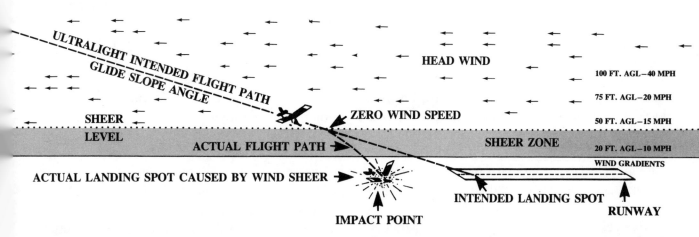

**WIND SHEER VALUE** – EQUAL TO TOTAL LOSS OF AIR SPEED THE MOMENT THE ULTRALIGHT PENETRATES THROUGH THE WIND SHEER LINE. RECOVERY TIME IS NEGATIVE BECAUSE OF LOW ALTITUDE AND REDUCED THRUST OF THE PROPELLER. THE ULTRALIGHT CRASHES. WHEN WIND SHEER IS ANTICIPATED, FULL OR CAREFULLY CONTROLLED THRUST IS MANDATORY FOR RECOVERY.

# TOW-TRAINING

**DAY 4**

The early morning drive out to Coronado airport was spent discussing, in detail, all of my flying events of yesterday. Bob and I carefully analyzed everything. This discussion helped to quiet the anxieties I had developed after the hard, nose-in landing of yesterday. Bob helped when he said that pilot trainees have a tendency to overcontrol, and that practice and good instruction can overcome the overcontrol tendency. I told him of the difficulty I had getting to sleep last night and he laughed, saying he had a similar experience the day before his solo. I chuckled to myself, thinking, I had to get through *this* day before my solo the next. I was determined to get "overcontrol under control."

My eyes fell on the flat topped Sandia Mountain off to the East of the airport. A slight fog obscured the valleys surrounding my mystical mountain's north and south sides. Today I could see it, I thought. Yesterday the weather completely obscured its view. I began to feel a lot better and was looking forward to the preflight and the day's lessons.

My preflight was all business. Nothing was overlooked. When I had finished I said to Kris; "She's ready to fly. But to be certain, Kris, did I miss anything?" I knew I hadn't but I always asked my instructor, for, I was the student. Kris moved over and depressed the choke on the Cuyuna and said: "No, I didn't remove anything. Warm up the engine and we'll get going. We have a lot of flying lessons today and I want to get them over with before the wind picks up later on. Right now she's blowing about 4 to 5 miles per hour out of the North West."

As soon as I started to taxi the Eagle, all thoughts and considerations I'd had earlier about flying immediately dissolved into the business of piloting the Eagle. I loved it. I shook myself loose and drove the airplane. A little throttle here and there to keep the roll constant but no quick bursts, just easy slow advances and retards while steering the nosewheel, and using the brake, with the left or right foot, depending on which way I was steering the Eagle. The Yellow taxi line stayed under the nose wheel all the way to the active runway where I stopped, looked in all directions for airplanes taxiing, taking off or landing.

All was clear. I slowly advanced the throttle to full and we scampered across the active runway. On the other side and up the incline, I pulled the throttle back to a little behind the cruise setting to maintain constant ground roll speed.

I taxied up to the end of the training runway, stopped beside Kris and Bob, killed the engine and hopped out of the seat. Bob pointed to the opposite end of the runway as he said. "Brian Allen is testing the two place Eagle with a sand bag as ballast in the passenger's seat." We watched as Brian made a smooth roll and takeoff. We held our eyes on him as he flew east towards the Sandia Mountains.

Kris had preflighted the Eagle and was taxiing to the end of the runway to perform his morning ritual of flying the Eagle before allowing me to use it.

I took many pictures of Kris taking off, flying and landing. He is an excellent pilot. I was more than pleased that he was my flight instructor.

On the first tow I was a little shaky but like the taxi, once the roll began I was all there, 100% of me.

Kris spoke to me through the radio and said that I should decide when to rotate for takeoff and that he would only speak to advise me of errors or to give instructions.

The tow line snapped taut and we went hurtling down the runway. I mentally counted the seconds; thousand one, thousand two, thousand three, thousand four. The wind was blowing across my face. The pressure was right for the 26 mph takeoff speed. Gently I pulled back on the stick. The nose lifted. I held the stick back. The mains broke away from the runway and we flew up to ten feet where I leveled off and held a straight and level course.

With smooth left and right moves of my wrist I swung the Eagle back and forth across the runway, making precise "S" turns while maintaining a constant altitude AGL. Kris released and when I anticipated touchdown was about to occur I flared and the mains started to spin on the surface before the nosewheel settled down to the roll-out.

My confidence and mastery of the controls improved with each run under tow. At times I flew to 15 to 20 feet AGL before leveling out. There, I gently depressed the left rudder pedal and noted the time lag before the Eagle's left wing fell back and put us into a turn instead of a yaw. It was exciting to refine the use of the rudder and spoilerons combined but it took sensitive handling and a positive attitude not to overcontrol.

On each tow I practiced the yaw swings back and forth over the runway and the "S" turns using the rudders and spoilerons.

On this fourth day of instruction I began to feel the cross wind on the takeoff roll and compensate for it by applying slight rudder either port or starboard, depending on the takeoff direction in relation to the angle of the crosswind. Once in the air, under tow, I still needed to use corrective control to compensate for the crosswind in order to hold her on a straight line over the runway and during the crossover maneuvers.

Two more runs and I felt great. My timing of the rotations, roundouts and landings were fine tuned. I looked forward to each moment of flying. On the last flight of the day everything went right. When we touched down and stopped after the roll out, I felt truly good. Brian Allen taxied up and asked if I'd like to take a turn around the patch in the two-place Eagle. I hefted the sand bag out, crawled in and buckled up. It was a marvelous flight. I felt proud and honored to have as my pilot the first man to fly across the English Channel under his own power. He was the engine. I scanned the horizon and the Earth below. It gave me a chance to see, from the air, the course I was going to solo over tomorrow. I took a lot of air to air pictures of Kris in my Eagle XL, flying the maneuvers I had been working on.

Today was the happiest taxi back to the hangar area I had experienced thus far. I had a smile on my face as I took the tip rudders off the Eagle and berthed her down for the night in the hangar. My last look at the Eagle XL had a thoughtful word attached. "Tomorrow, we fly together, unattached, under your own power." I ran my hand gently along the Eagle's wing.

That night Bob took me to dinner and we had a great evening together in Albuquerque. I had no trouble sleeping that night.

1. **TOW CABLE TIGHTENS.** Ground roll to 26 mph.

2. **26 MPH.** Slight back stick rotation. Liftoff.

5. **STARBOARD RIGHT SPOILERON** deflection. Roll right.

6. **STARBOARD TIP RUDDER** deflection (slight). Yaw right and cross centerline on runway.

9. **ALL CONTROLS NEUTRAL.** Wings level. Airspeed decreasing. Slight spoileron deflections—port and starboard to maintain level wings in the crosswind.

10. **MAINTAIN LEVEL WINGS.** bring Eagle over centerline by using delicate control inputs.

13. **TOUCHDOWN.** Roll-out. Crosswind trying to lift port wing.

14. **TOW CABLE RELEASED.** Roll-out to stop.

3. **GAIN ALTITUDE.** Compensate for cross-wind.

4. **PORT (LEFT) SPOILERON** deflection. Roll left.

7. **CANARD ROTATION DOWN.** Gain altitude. Rudder port. Yaw left across runway centerline.

8. **SLIGHT STARBOARD RUDDER** and spoileron deflection. Right coordinated "S" turn across runway centerline.

11. **AIRSPEED DECREASING.** Losing altitude. Compensate for crosswind with sensitive spoileron deflections.

12. **LINE-UP ON RUNWAY** centerline while continually compensating for the cross winds.

15. **KRIS UN-HOOKS TOW** cables as Brian Allen takes-off in the two-place Eagle XL.

16. **TWO-PLACE EAGLE XL** with Brian Allen seated in the cockpit while Kris and Bob look on.

# PILOT BRIEFING
## STABILITY

## PILOT INDUCED OSCILLATION (PIO)

When a pilot rotates the stick back quickly and the ultralight pitches the nose up severely and then the pilot rotates the stick forward quickly to compensate and the nose pitches down, then the pilot pulls back to correct the down-nose attitude, the ultralight can severely oscillate up and down to the point where the ultralight becomes totally out of control. When high airspeeds created by quick throttle advances and hard control stick forces are strong enough the ultralight airframe can fail. The cause of the failure is known as "Pilot Induced Oscillation – PIO." The solution is smooth (stick and throttle) handling, proper flight training, a thorough understanding of flight dynamics and staying within the manufacturers' flight envelope (limit) of the Ultralight.

## STABILITY

STABILITY—State or quality of being stable. An airplane's ability to return, stay at or move from its original flight direction after being dislodged from its original flight direction by an outside force.

LONGITUDINAL STABILITY—The airplane will not nose up or down from the prime flight direction but will remain stable along its longitudinal axis unless re-directed by pilot control.

LATERAL STABILITY—Wing dihedral is a prime design factor in creating lateral stability. Wingtips will remain in their prime wing attitude unless displaced by external forces such as a gust or pilot instituted control.

VERTICAL DIRECTIONAL STABILITY—The airplane maintains a straight flight line by the force of the relative wind reacting against the vertical tail surfaces.

NEUTRAL STABILITY—The airplane stays in its intended flight attitude.

NEUTRAL STATIC STABILITY—The airplane stays in its prime flight attitude without pilot control.

POSITIVE DYNAMIC STABILITY—After a series of diminishing oscillations the airplane returns to its prime flight direction without pilot control.

NEUTRAL DYNAMIC STABILITY—An airplane that swings back and through the intended flight direction and continues to oscillate back and forth, without decreasing the oscillations.

POSITIVE STATIC STABILITY—When an airplane, without pilot control, rapidly returns to its original flight direction after being forced off course by outside forces. Wing dihedral and CG location are vital factors in designing Positive Static Stability into an airplane.

NEGATIVE STABILITY—The airplane continues on from the prime flight direction unless stopped by pilot control.

NEGATIVE DYNAMIC STABILITY—An airplane that deviates from the prime flight direction with a dramatic increase in wild oscillations that grow in intensity and requires violent pilot control to prevent crashing.

# FLIGHT CONTROL

FLIGHT PLAN—A wise, safe pilot always preflights each flight before taking off. On cross-country flights, have the proper sectional maps and FSS (Flight Service Station) numbers on hand for up-to-date weather and route flight conditions.

PREFLIGHT—Inspect and feel the airplane. Listen to the tune, sounds and vibrations during run-up, take-off, while in flight and landing. Listen for any variations not normally present so you can anticipate and correct a malfunction before it becomes a crisis.

TAKEOFF—Thrust from the engine driven propeller moves the airplane's wing airfoil into the relative wind at a speed that overcomes the force of gravity and drag with a lift force and the airplane rises into the air. On the Eagle XL this lifting force occurs at a speed of 26 mph.

Once off the ground, airspeed is increased by lowering the angle of attack to level flight, increasing throttle until a climb/glide speed of 35-37 mph is attained.

When 35-37 mph is reached in level flight, increase the angle of attack to climb to the selected AGL attitude.

Throttle back to maintain a cruise speed of 35-37 mph.

**TAKE-OFF DISTANCE—75 to 100ft.**

**FULL THROTTLE—** Stick Neutral

**GROUND ROLL SPEED** 26 MPH—Stick Neutral

**GENTLE ROTATE STICK BACK AND HOLD UNTIL LIFT OFF—** Airspeed 30-35 MPH

**HOLD STICK BACK FOR CLIMB-OUT—** To Maintain Airspeed 30-35 MPH

**AT 4 FT. AGL—** Level-off until 35 MPH airspeed is reached. Then rotate back on the stick for the climb-out.

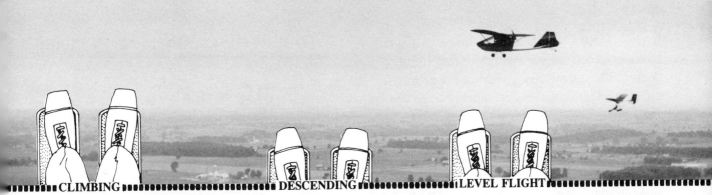

**CLIMBING ........ DESCENDING ........ LEVEL FLIGHT**

**IN OPEN COCKPIT ULTRALIGHTS THE TOES** of your feet are excellent marks for judging the relationship of the nose of the aircraft to the horizon. Toes above horizon—climbing. Toes below horizon—descending. On the horizon—level flight. Always scan the airspeed indicator to maintain constant airspeed.

To maintain straight and level flight at a constant altitude the control stick is used (Port—Left; Starboard—Right) to first level the wings with the earth. If the aircraft nose is high over the horizon, slight *forward* rotation of the stick will lower the nose. Should the nose be pointing beneath the horizon, slight *back* rotation of the stick raises the nose to the desired cruise level.

To increase airspeed while in level flight—advance the throttle. The additional thrust produced by the propeller will also increase the wings' lift. Slight forward pressure on the stick will hold the nose in a down (to compensate for the additional lift) *altitude* which will maintain straight and level flying. When the throttle is *retarded* the wing loses lift and the wing must be set at a higher angle of attack to maintain level flight. This is accomplished by slight back rotation of the stick. Now the nose points slightly above the horizon while maintaining level flight. These maneuvers are called "Attitude Flying."

# TURNS

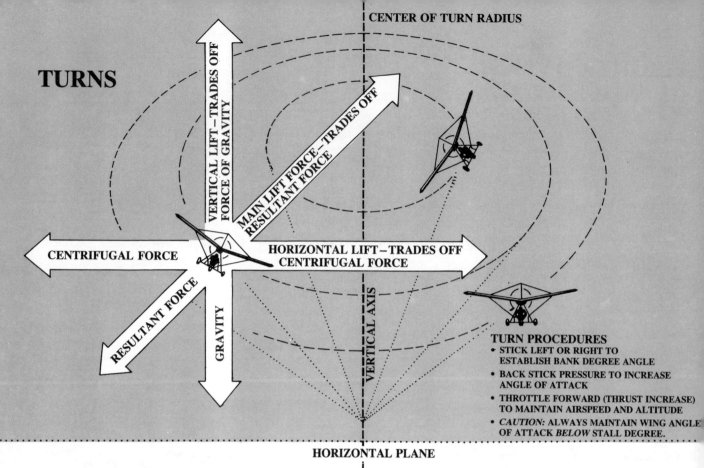

**TURN PROCEDURES**
- STICK LEFT OR RIGHT TO ESTABLISH BANK DEGREE ANGLE
- BACK STICK PRESSURE TO INCREASE ANGLE OF ATTACK
- THROTTLE FORWARD (THRUST INCREASE) TO MAINTAIN AIRSPEED AND ALTITUDE
- *CAUTION:* ALWAYS MAINTAIN WING ANGLE OF ATTACK *BELOW* STALL DEGREE.

**TURN RADIUS BANK ANGLES**

As the turn radius decreases, more thrust and wing airfoil angle of attack is required to maintain a constant altitude and to overcome centrifugal (G) forces and the increase in drag (D).

**LEVEL FLIGHT** — To return to level flight, bring the stick to neutral which will set the nose on the horizon and complete the round-out maneuver.

BANK TURNS — When the wing of an airplane is inclined to one side or another during flight, a turn occurs. During this maneuver four opposing forces are in balance. They are: Centrifugal Force, The Turning Force, Vertical Lift and the Force of Gravity.

When Centrifugal Force and the gravitational forces are combined they are called the Resultant Force.

As the bank angle is increased, more lift is required to hold the airplane in level flight. In addition, the Resultant force also increases and places a higher (G) load on the aircraft's wings and tail surfaces.

On the Eagle XL, the critical (Vne) bank angle is 60°. These bank angles are determined by the angle of the wing as it relates to the horizon.

In turns it is necessary to increase thrust (throttle up) and the angle of attack (stick back, pitch up) of the wings airfoil. The result is Lift's vertical and the horizontal forces overpower (trade-off) the gravitational and centrifugal forces that if left un-checked would bring the airplane to earth.

## FORCES ON AN AIRPLANE IN TURNS

| Bank Angles in Degrees to the Horizon | Center of Gravity Forces | Stall Speed % Increase |
|---|---|---|
| 0° | 0    0G | 0 |
| 10° | 1/4 | 20% |
| 15° | 1/2 | 25% |
| 30° | 1 G | 60% |
| 40% | 1-1/2 G | 85% |
| 50% | 1-3/4 G | 100% |
| 60% | 2 G | 120% |

--------------- Vne Never exceed bank angle ---------------
S T A L L

TURNS — Ultralight turns can be activated several ways. Slips and Skids result when the airplane rotates on its longitudinal axis without a loss of altitude.

ALTITUDE — Is safety — if engine fails the pilot has a chance to find a suitable landing site. If a structural failure occurs there is a safety margin to use the parachute.

LANDING — A Glide Speed of 35-50 mph is maintained as the airplane is pitched down. When the landing zone is approached, Thrust is reduced by throttling back while maintaining an angle of descent speed of 35-37 mph (even with the engine off). Air speed is critical, when landing, as a too-low and slow, airspeed below 35 mph, can cause a serious accident should the airplane fly into a wind sheer, or gust. It is always better to maintain a higher air speed when landing than a lower air speed. Control is the magic flying word. At a low air speed there is little or no room to correct, or adjust, errors.

A few feet AGL, throttle all the way back and round out the descent to level flight and hold off landing until the air speed reduces. The angle of attack is gently increased. Speed decreases and as it does, come back on the stick just before the main landing gear wheels touch down, and bring the airfoil (wing) into a flair.

When the main wheels make contact, additional air speed is lost and the nose wheel drops down and the airplane rolls to a stop.

Proceed with the taxi procedure to move off the active runway onto the taxi access runway.

When you get to the parking location be certain to tie-down. Many airplanes have been lost because the pilot neglected to tie-down the airplane after landing.

## WAKE TURBULENCE

All airplanes when taking-off or landing produce wake turbulence. Air flowing off the wingtips, from beneath, and over the wing produces rapidly rotating vortices of air that trails behind the airplane. Heavy, large, slow airplanes produce the highest intensity vortices within the wake turbulence.

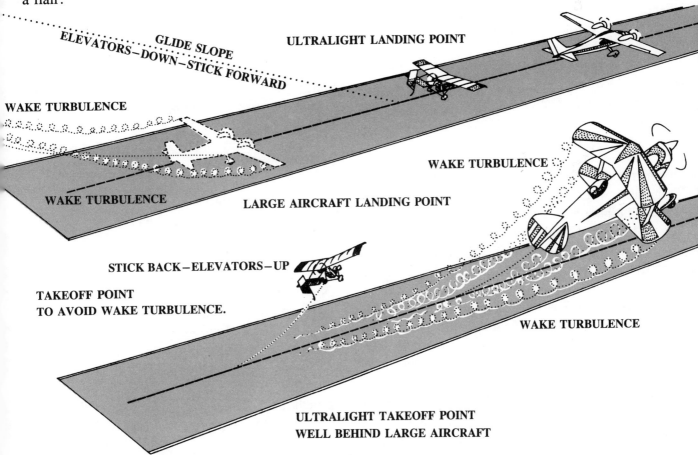

**TO AVOID "WAKE TURBULENCE"** behind a large plane's landing final, always land your ultralight *well ahead* of the exact spot where the larger plane touched down. Otherwise your ultralight can be tossed about like a leaf in a windstorm. When taking-off from behind a large aircraft you should plan your liftoff point to be far behind the exact spot where the larger aircraft made its take-off rotation. Otherwise, your ultralight could be twisted in the air like a pretzel. A recent report stated that an ultralight was completely destroyed by the wake turbulence of a jet transport one-half-hour *after* it had taken off. Beware of wake turbulence. It can hurt you and your ultralight.

# SOLO EAGLE XL

**DAY 5**

The day for earning the wings of an eagle was upon me. My preflight inspection of the Eagle XL in which I was going to solo had been thorough, and all was in readiness. Kris Williams, my flight instructor, had removed no critical safety rings or pins from the bird as he had done before to see if I would miss a vital part in my inspection. Still, I turned to him.

"She looks good to me. Did I miss anything?"

Kris shook his head. He knew the excitement boiling in my blood. He'd seen many fledglings on the mornings of their solos. He handed me my helmet and hooked up the two-way radio.

"Hop in."

As Kris and Bob worked over me, they reminded me of eagles preening their fledglings for their first flight. Safety harness over my shoulders, velcro down the chest strap to hold it tight, helmet strap secure and fastened, seat belt buckle snapped shut and tightened. Kris finally plugged in the radio jack and spoke into the hand transmitter.

"Are you receiving me?"

I rotated the canard wing up and down, the standard signal received procedure, and smiled as he slipped onto the station wagon tailgate. His voice came in clearly through the headset earphones.

"Start the engine and follow us."

Quickly I double-check the fuel lever, making certain it's on. I reach back, depress the choke, turn the ignition switch on, grasp the overhead starter handle with both hands and shout,

"Clear prop!" and yank down hard.

The cold 30 horse power Cuyuna engine gave a short chug as it turns over silently. I pull again. Another deep throated cough. Still silent. Bob stops the wagon and Kris slips off the tailgate and starts walking toward me. I give it my all this time, and with a roar the powerful Cuyuna springs to life, shattering the early morning silence with an angry mechanical snarl.

I set the trottle at idle and check the controls. A wisp of blue smoke curls quickly away in the prop blast. My right hand moves the stick right and left. The roll-controlling top wing spoilers flutter up and down smoothly. My eyes fix on the canard above and ahead of me. First the leading edge slats, which direct air flow over the canard to increase lift, making certain they were attached tightly. Then the canard elevator system. I rotate the canard up then down. All linkages were smooth as silk. My right foot pushes forward and the right tip rudder swings smartly in as the nose wheel turns squeakily to the right on the tarmac strip. Left tip rudder the same, and I take my foot from the brake post between the pedals, increase the throttle, and the Eagle begins to move forward.

Center the nose of the Eagle on the station wagon; more throttle and we bounce down the taxiway, eyes constantly on the wing tips to keep clear of parked cars, planes, gas pumps and hangar doors. Keep the nose wheel on the yellow line as we make a sharp turn up to the main runway. More power on for this uphill climb. Throttle back and press my left foot on the brake post to keep from running into Kris watching me from the station wagon tailgate as they stop before crossing the main runway.

"A plane is taking off, Rick. As soon as he's clear follow us right across the runway." A climbing Cessna speeds past. Bob rumbles the wagon across and the Eagle follows smartly at three-quarters throttle.

"Rick, when we get to the end of the runway stop and kill the engine. I want to take it up for a test flight." Soon we were there. I stopped, unhooked, and crawled out.

I watch carefully making mental notes as Kris executes his preflight with smooth proficiency, cranks the bird up, and was quickly airborne in the early morning desert sunrise.

Bob strode over to me as I loaded my camera. I handed the Canon to him.

"It's on 'program.' All you have to do is focus and shoot. All automatic. I want *this* solo documented. Do you mind?"

He took the camera, looped the strap around his neck, and snapped a picture of Kris on his final landing approach.

"Simple enough. I'll be glad to, Rick."

While Bob busied himself removing the tow cables from the back of the station wagon, my mind whirled.

"I thought I was going to solo today. Maybe they think I'm not ready yet." I dreaded the feeling of failure. I'd been here six days. Perhaps soloing needs more training than I'd had.

My mental considerations were interrupted as Kris taxied up to us, cut the engine and got out.

"Rick, before you solo we're going to do a couple

of fast tow flights, just to familiarize you with takeoff and landings." A smile crossed his face as he said,

"You're ready to fly, but—" he looked northwest across the desert at the fluttering wind sock, "There's a slight crosswind blowing. I'd like you to make a couple of takeoffs and landings under tow before you solo, so we'll hook up now and give it a go."

Within minutes I was trailing behind the wagon, and lifting eight or nine feet into the air. Kris's voice came over the radio.

"Alright now, Rick. Use the spoilers only, and make a few crossovers, first left and then right. Stay within a few feet of the centerline on each crossover."

I gently move the stick to the right. The spoileron flutters up, then down as I center the stick. The Eagle swings gently to port. A little port on the stick brings her across the white centerline. Before crossing the Eagle to the precise three-to-four-feet left of the centerline, I ease the stick just a touch to port.

"No overcontrolling this bird today, Rick," I think to myself. The Eagle responds smoothly—

"A little too far, Rick!" (Like Kris said. I was not three-to-four-feet, more like nine or ten feet, over the center line.) We're over the side of the runway, I give her a little right stick, and we sail smoothly back across the line.

As I am making the next right turn and the Eagle crosses the centerline, Kris releases the tow. I bring the stick just a touch to the left as the bird descends. A little right spoileron, and the Eagle's wheels are about two feet above the runway. Back on the stick, flare and the rear two wheels kiss the tarmac. Canard forward and the nose wheel touches. I am down. The earphones crackle,

"Beautiful landing, Rick! Now roll up to the end of the runway and we'll do another fast taxi and try it again."

I braked the Eagle to a stop and took a deep breath. Kris was beside me.

"That was great, but you flew too far to the right. On this tow, try to keep it only a foot or two on either side of the centerline. We'll go a little faster this time, so take her up fifteen feet before you start the turns. I'll only speak over the phones if I think you're not doing it right. Now, on the run down the strip, you decide when you have enough airspeed to lift off. You make the decision."

These fast taxies to the end of the runway really

**A CESSNA** takes off in the early morning fog.

**KRIS TEST FLYS** the Eagle Xl before training exercises.

**AIRBORNE UNDER TOW.** Delicate but positive control inputs to compensate for cross winds.

**PORT SPOILERON** (left) deflection to fly the Eagle XL over the centerline.

**A CROSS WIND** lifts the port wing on takeoff.

**AS HIGHER ALTITUDE** is reached, the cross wind become stronger.

**PORT SPOILERON** deflection brings the Eagle's wing level with the horizon.

**PORT RUDDER DEFLECTION** yaws the Eagle XL towards the centerline without loosing altitude.

set you on edge, I think as the cable begins to snake out in front of me, grows taut, then snaps as it tightens up. Bob accelerates the wagon. I clench my teeth, determined to keep that nosewheel right on the centerline for the fast tow back to the end of the runway. I hold it right on until Kris's release, then I taxi up and stop. Kris is quickly beside me.

"Remember now, when you get up in the air stay just two or three feet to either side of the centerline on the crossovers." He glances at the wind sock.

"It's freshened up a bit, so compensate for the crosswind."

I nod but say nothing. As Kris walks back to the tailgate, I say to myself,

"Dammit, Rick, get it right this time!"

The tow rope tightens and the Eagle starts briskly down the runway, faster than I have ever been towed before. At the precise instant I feel the speed is enough, rotate back on the stick. The canard responds, and "pop!", she lifts into the air. I hold the stick back until the tow rope is angling down sharply to the tailgate a good fifteen feet below. The phones vibrate.

"OK, Rick, start your crossovers."

A little stick to the left this time, into the crosswind. The Eagle gently swings a mere foot off the centerline. Instantly I flick the stick to the right. The Eagle darts three-to-five feet across the centerline.

"It's that crosswind," I think. "No, turkey, it's that time lag and the crosswind. Compensate for it on the next crossover." Left stick again. Then quickly back to center. The Eagle bucks the crosswind and slowly moves to the two-foot measure beyond the centerline. The end of the runway is coming up fast. Gotta hurry before Kris releases. I flick my wrist to the right and instantly back to neutral. "I got the time lag problem that time I think." I flick my wrist to the right and instantly back to neutral. The Eagle swings a neat two feet to the right. The moment I'm flying over the centerline, Kris releases. Stick just a wrist nod forward and the runway is coming up. My wheels are a foot off the runway, rotate back on the stick, the canard comes down, the Eagle flares, rear wheels touch, canard up, nosewheel down and roll to the end of the runway.

Kris and Bob come up to me as I sit quietly in the seat. Kris spoke.

"That was perfect, Rick!" Bob leaned forward.

"Well, Rick, you're ready for solo."

I unsnap the safety harness, clamber out and stretch my legs. I am turning into a computer. I repeat the preflight protocol after the tows, checking everything. Study the clouds. Look at the wind sock—it's tail down but fluttering. The crosswind is about the same. Kris and Bob unfasten the tow line, coil and stow them in the station wagon. They do this silently, giving me a chance to collect my thoughts.

The feeling coursing through me must be similar to that of a ski jumper poised at the start of his first jump, or that of a novice diver about to go off the ten meter platform into the pool. But the apprehension is muted by recalling the careful training and confidence Kris and Bob have given me.

"No use waiting around anymore," I thought. I slip on my helmet, make one last final preflight checkout, climb into the cockpit, and hook up the safety harness. Bob and Kris walk over to me.

"You look like you're ready."

I smile, "Ready as I'll ever be. Let's go for it."

Kris depresses the choke, yanks hard on the starter, and the Cuyuna breathes fire as Bob grasps both of the down tube framing members above me. Kris runs the engine up to full throttle, and the Eagle vibrates with 230 pounds of thrust pushing against Bob. Kris returns the throttle to idle. Both Kris and Bob give me a thumbs-up as they walk back to the wagon. I aim the Eagle towards the far end of the runway.

The moment of final decision: To fly or not to fly. I fly.

"Here we go!" I advance the throttle. The Cuyuna whirls the hardwood prop at taxi speed. I glance at my thirty-five foot span of wing bobbing up and down as we bounce down the hard tarmac, nosewheel tracking perfectly on the white centerline—that pesky wheel had been a problem on ground tows. See, too, the brown clumps of grass bordering the narrow runway—serious problems there two days before during a fast tow. A chill darts through my body: "You must keep that nosewheel right on the white line on *this* takeoff, Rick, old boy!"

The end of the runway looms ahead. Brake, a little power, and the Eagle taxies to the side of the strip. Putting my feet on the tarmac, I sidestep the bird to the centerline.

Kris's voice crackles on the headphones,

"If you're receiving me, rotate the canard."

**TOUCHDOWN** on the centerline.

**BACK INTO THE AIR.** Starboard (right) spoileron deflection prevents the cross wind (blowing from the right) from lifting the starboard wingtip.

**JUST BEFORE TOUCHDOWN** the cross wind raises the starboard wingtip.

**SLIGHT OVER-CORRECTION** for the cross wind dips the starboard wingtip on touchdown.

I swallow hard and rotate the canard up and down. His voice is thin over the radio.

"Remember, Rick, keep the nose straight down on the white centerline, and when you throttle up to full power and reach takeoff speed, rotate the canard slowly but positively back. When you reach two-to-three hundred feet altitude, turn right down along the fence-line beyond the end of the runway, then proceed to the small hillocks at the foothills of the mountains. Then turn right again parallel to the runway and make a pass over it at altitude. I'll be on the radio all the time, but I'll only talk if I think you're in trouble. And Rick, this is from both Bob and myself, have a good flight!"

I wiggle the canard in acknowledgement, lean back in the seat, suck in a deep breath, control the flowing adrenalin, relax, and work the controls one by one. Depress right foot pedal and the right tip rudder swings in, and returns with a gentle thump as I release the pressure; similarly with the left. Stick left, and the shadow of the wing spoiler shifts on the wing fabric.

The canard smoothly rises and falls as I fore and aft the stick. I look back at the pulsing Cuyuna and inadvertently brush my helmet against the airframe. The intense vibration rattles and buzzes my head like a nest of angry hornets. I adjust my helmet, sunglasses, and seat belt, and carefully tuck in the red and white bandana, remembering the story of the hapless pilot whose scarf blew off and tangled the prop, causing a quick and unintended emergency landing.

I am sweating in the cool desert morning. This is the end—or the beginning—of the line. No more negative thoughts about flying. I was properly trained. Kris and Bob, my instructors, are as concerned as I am. Bob has the Canon with a full load of film. If anything happens, he'll have a record. My thoughts flash as an inner loud voice shouts in my head.

"Carrier, what in hell are you doing?"

"This is your solo, you idiot! That's why you are here. Let's go for it!"

My left hand slowly inches the throttle forward. The Cuyuna growls. I hold my foot on the brake post between the wingtip rudder pedals and add more throttle. The Eagle trembles as the fuel mixture surges through the carburetor, and the prop slaps the air with a loud, high-pitched rapping that chops through my earphones. I release the brake and 230 pounds of thrust slam me back in the bucket seat and hurtles us down the runway. Eyes and nosewheel glued to the white centerline.

"Keep it straight, Rick," my inner voice reprimands. "No off the runway and into the scrub grass today." The centerline whips under my wheels. One second, two seconds, three seconds, the white center dashes blur. Eyes fixed on the end of the runway, four seconds.

I rotate gently but firmly full back on the control stick. The nose rises sharply and the Eagle leaps from the runway! I hold my breath. We are pointing what seems like straight up at the clouds. The Cuyuna is in full roaring voice. Higher we go.

My nerves tingle. Adrenalin pumps through my heart. That moment so often dreamt of is alive in reality. I am separated from the earth and climbing fast, free of the demanding force of gravity, conquered by the powerful thrust of the thirty horsepower Cuyuna and its fifty-four-inch hardwood propeller.

I take a quick look over my left shoulder at Kris and Bob below me, shrinking in size beside the wagon on the end of the runway.

No time for sight-seeing. Concentrate on the horizon. Keep the wings level. Hold the throttle wide open. Gain altitude, altitude . . . altitude. I recall Bob's comments the day before when another fledgling pilot gained his wings. At takeoff, he feared altitude.

**YOUR MIND** whirls faster than the propeller on the taxi out to the end of the runway for your solo flight. Items: "Have I preflighted everything? Can I make it? Is my insurance paid up? Safety Harness secure?" The answer is yes to all. Now is the time to fly.

He had leveled off at less than a hundred feet, and started turns and banks over wires and fences while keeping the engine wide open. Bob's words rang in my ears.

"He's much too low and at full throttle. If the engine quits, he has no room to maneuver or look for a suitable landing site. He'd have to go right in and land wherever he could and that's bad. Altitude, Rick. Remember on takeoff *gain altitude*, plenty of it. It'll give you a chance to think and react if anything should happen—and things do happen if you fly enough. So be prepared. Altitude will give you that edge for the safety margin you'll need."

With a quick scan of the altimeter I realized I was above the 300 ft AGL flight restrictions for ultralight training flights. Easing the stick forward quickly brought the Eagle XL to the required ceiling where I leveled off and scanned the sky and earth for other aircraft and landmarks.

The dirt road where I'm to make a ninety degree right turn is not far ahead. Kris's voice vibrates in the earphones, his words mixed with the static caused by the engine's ignition system, audible, but requiring close attention.

"Rick, you're doing fine. Make a ninety degree turn right at the dirt road, fly down to the flat-top hill on your right and make another ninety degree turn and fly parallel to the airport. I'll stay off the air unless you get in trouble. Fly the plane a little."

All my training now comes into play. Those crossovers that under tow were so hard now seem almost too smooth, too easy. A soft twist of the wrist to the right, hold, and the Eagle banks into a gradual turn towards the flat-top hill straight ahead. Just before I'm lined up with the flat-topped hill I bring the stick back to center, remembering the response lagtime. When I'm on course I take a moment to look around, relax the intense concentration just a bit. I'm still climbing. I squeeze the throttle grip in my left hand and ease it back until the engine tunes down to a steady drone. No chainsaw growl, just a fine vibrant purr.

The surface of the earth peels away beneath me. A glance at the horizon, and the slope-shouldered Sandia Mountains loom majestically ahead. That flat-topped mountain that had held my attention from the ground is straight ahead. I'd heard that the native American Indians had religious ceremonies at a secret spot on these mountains, celebrations to the spirits.

**AT FULL THROTTLE** the thrust from the propeller drills the Eagle up into the sky.

**STICK FORWARD.** Level-off. Pick up airspeed to 40 mph. Ease the throttle back to a cruise setting.

**STICK BACK.** Gain altitude and turn North to the flight pattern boundaries that parallel the Sandia Mountains off to starboard.

**STARBOARD FOOT PEDAL** depressed. Right tip rudder rotates inboard and the Eagle smoothly yaws right. Hold for 90° turn, then pedal back to neutral.

**COCKPIT VIEW** of the Sandia Mountains looking east from 300ft AGL. The Flat-topped mountain is to the right in the photograph above. BELOW: Slight starboard spoileron deflection up and gentle starboard tip rudder deflection inboard turns the Eagle into the proper heading for the downwind leg of the landing approach.

No one knows the precise place. Deep down inside, I had a mystical feeling on several occasions that the flat-topped mountain was the Indians' sacred ceremonial site.

I had vowed to fly over it at some future date and see if that feeling was confirmed, for I felt at this moment that, there on that mountaintop the Indians worshipped the flight of the eagle, as symbolic of the spirit of their departed ancestors. I was one with them now. I dipped the nose of my Eagle down, and up to the heavens, acknowledging the thought.

Kris spoke over the headset.

"You're far enough down on the pattern. Now make a ninety degree to your right and fly parallel to the field, and when you are beyond the end of the runway, make a one hundred and eighty and then a practice landing in the air over the runway. Then go 'round and when you are ready, make your final, and land."

Back to the business at hand. Concentrating intensely, I ease the stick to the right. The Eagle banks more steeply this time. I use the tip rudder with the spoileron. Just before the nose is parallel to the runway, I bring the stick a little left off central and then back to center. The Eagle responds with beautiful clarity. We are aiming directly at the tall stand of trees beyond the end of the runway. From the ground it was impossible to see what is on the other side of the trees. From up here, the question is answered: A cemetery!

I shudder, feeling I can see the hands of the dead waving above the tombstones at me, their ghoulish faces grinning, hoping I would make a fatal mistake and join them in oblivion.

"No trips to the boneyard for me," I thought. "My landing, that's real, Rick. Forget idle thoughts. I have a maneuver to execute." Thinking had taken time, and time is supremely important in flight.

I was well beyond the end of the runway. Kris's voice reverberated in my ears.

"You've gone too far, Rick. Make that one-eighty and come back over the field."

I twist my wrist to starboard and deploy the starboard tip rudder with my right foot. It is a hard, steep turn. The Eagle digs the right wing tip into the air, and sweeps into a severe, right-banking turn. The quickness and extreme angle of the turn catch me up short. I clench my teeth and increase throttle to hold the nose up. The Eagle shudders with the sudden increase of the Cuyuna's rpms. The nose rams above the horizon as I pull back on the stick. "Hold it tight into the turn, Rick, glue your eyes on the end of the runway far below in front of us and scan the area for other aircraft." All clear. As soon as I am lined up on the runway approach I bring the stick to center and ease it forward. The Eagle lowers its nose; the wind blows hard on my face as we dive. Teeth gritted, I hold her tight in the dive. I am still too high. "Come down to at least one hundred and fifty feet for the flyover, Rick."

I count: One second, two seconds, three seconds—with the thought of those skeletons in the graveyard behind, reaching for my body, puts a thought in my head. "Don't give them a chance, Rick." I look at the altimeter. I am two hundred feet AGL—enough altitude for an experience, I think. I glance back at the cemetery. The hands are beneath in their graves.

"It's OK now, Rick. Just use your judgment. If you want to try out a flying maneuver, go ahead." My mind waited for an answer. It came.

"Just do what Kris said to do. That's all."

I looked closely at the earth below. It was crawling by. The runway was fast approaching. What was down there that made me look so hard? It was the dimensions, the angles. I had never looked at the Earth like this before. Flying as a passenger in a conventional plane gives you a sense of security. You trust the pilot with your life while you read a magazine, sip a drink, while you engage in delightful conversation with a widowed heiress of a millionaire, who asks for your address and gives you hers. You have made plans for tomorrow, while flying 35,000 feet in the sky. That faith and security you feel is in the ability of your pilot, and is thus very important.

Right now I am the pilot in command. The responsibility overwhelms me. It is no longer a dream. This is the reality. I am completely, absolutely, eternally, responsible for the safe return of my body to earth. Should I crash, the long hands of death in the cemetery here will await my flesh. Then my spirit will fly to the top of that mountain to my east, and I shall fly this desert for the rest of my spiritual existence.

I choose having a beer with Bob at the bar of the Marriott House in Albuquerque, and the scents of the beautiful women there.

—Four seconds. Hold that stick forward! No hesitation now, Carrier. The vibrations increase dra-

**INCREASE THROTTLE**—The additional thrust creates more lift and the nose pitches up. Starboard tip rudder yaws the Eagle into a right turn into the base leg of the landing approach.

**THROTTLE BACK TO CRUISE** setting all controls neutral. Wings level with horizon. The cemetary is at the top left of the picture above. BELOW: Enter the turn into the final leg off the landing pattern.

**STICK FORWARD**—Hold nose down on final practice-landing approach—200ft AGL.

**STICK BACK**—Round out. 150ft AGL.

**RE-ENTRY** into the landing pattern. BELOW: Turning 45° into the downwind leg.

matically as the wind blows against my face. Five seconds. The air rushing behind my sunglasses makes me blink to clear my tears. Six seconds. Faster down.

At the moment I know Kris will interrupt, stick back, round out and throttle back for a smooth fly over the field. I'd dropped 150 feet in the dive. It was like being on a roller coaster, at the bottom of the run. I feel the G-force of the pullout. It's not that dramatic, not the gut feel of a Helldiver bottoming out. It was a simple forty-five degree pitch down slope. But for me, I felt like a flyer on a diving run against formidable odds I had seen happen so many times in flying movies. I was John Wayne in "Flying Leathernecks"; Errol Flynn in "Dawn Patrol"; James Cagney flying bush pilot in Alaska. All of them were rolled up in me as I flew over that landing strip, deep in the desert dawn in New Mexico.

Kris's instructions rang in my ears. "Try a high altitude practice approach and landing. Get familiar with landing pattern entry procedures before you attempt the actual landing." I rocked the Eagle's wings up and down to acknowledge receipt of message. My eyes scanned the airport.

The only activity was a twin Beechcraft on final approach far to my right. I held a steady course as I watched the airplane make a smooth touchdown and roll-out. I vowed to do the same on my solo. I reviewed the landing procedures in my head, just as Kris had outlined. The entry window to the Down Wind Leg was approaching.

My altitude was 150 feet AGL. I eased the stick to starboard and the Eagle smoothly responded by tilting her wing to the right as we entered the Downwind Leg at the prescribed 45° entry angle.

The eastern end of the runway was about a thousand yards ahead and about a football field to my right. The turn and pattern entry procedures had eaten up fifty feet of altitude. The altimeter clocked 100ft AGL. Easing the stick to port I lined the Eagle parallel to the runway and a few quick twists of the stick leveled the wing with the horizon.

My eyes scanned everything. Horizon, flying wires, tip rudders, the entire area around the airport, ahead, above, to each side, below and behind. All clear. I eased the throttle back to 3/4 setting, lowering the Cuyunas thrust and rpm's.

As I approached the southern end of the runway I picked the spot on the runway surface where I would make touchdown if the actual landing was made. I was heading Southwest flying parallel to the runway which lay about one hundred yards off my starboard wing. As the southern end of the runway passed, I glanced back over my right shoulder to recheck the entire area. It only took a few seconds. All was clear. The time to enter the Base Leg of the landing pattern was now.

My right foot gently depressed the starboard tip rudder pedal and my wrist eased the stick starboard. The Eagle banked right sharply in a coordinated turn. The nose went down. That was OK but the swift turn caught my breath. With positive control I brought the stick back to center and depressed the port tip rudder for just a moment then brought all controls back to neutral. I was in the Base Leg with the wings level with the horizon but I was losing altitude. The altimeter read, 75ft AGL. I held the nose up just a few degrees above the horizon and the clock held on 75ft AGL.

Calculating the 90° turn into the Final Leg required anticipating the exact moment to execute the maneuver. It was vital to come out of the Final Turn "Dead Center" over the runway without losing altitude. The moment came.

I eased the throttle forward a touch and gave easy starboard stick and rudder deflection with a tender pull back on the stick to hold the nose on an imaginary line just above the horizon. A split second before lining up center on the runway I brought all controls back to neutral. The time lag was just right. We were aiming straight down the runway, over the centerline. It only took a second to notice that the crosswind was forcing the Eagle off to starboard. We were heading towards the right side of the runway. Tender, slight port tip rudder deflection and a quick touch of port spoileron and the Eagle crabbed back towards the centerline. The altitude was holding at 75ft AGL.

I can practice a little, with the safety of 75 feet of altitude. I pull the throttle back to idle. Nose down to fifty feet AGL. Pull back and level out one quarter of the way down the runway. I picture landing. At the moment I feel that if we were close to the ground touchdown would be imminent, pull back on the stick. Flare. OK, Rick, no stall now. Stick forward. Throttle forward. The Cuyuna roars awake. Nose up and we are gaining altitude as I fly over Kris and Bob waving below. Power on full.

**STARBOARD.** Spoileron deflection to level wing.

**ENTERING** a coordinated turn — starboard tip rudder and spoileron deflection.

**INTO THE BASE LEG** — 75ft AGL. Below into the final leg of practice landing — 75ft AGL. Then throttle up for fly-over.

**CLIMB-OUT** to 200ft AGL. Then starboard turn into downward leg of the landing pattern.

**ENTER DOWNWIND LEG.**

**ROUND-OUT INTO FINAL LEG. BELOW:** Port spoileron up to level the wing.

Nose high over the horizon. Hold the climb till 200 feet AGL. Then turn parallel the runway. Power back to mid-cruise, runway gliding slowly beside me. Power back a little more. Aim for the trees bordering the cemetery. Down to one hundred feet. "Time to enter the base leg," I thought. Wrist to starboard. Spoileron flutters up. Tip rudder swings out as I depress right pedal. The bank angle increases. A surge of excitement as the horizon tips alarmingly, I'm too high. Throttle back a little. Stick all the way forward. The Eagle's nose pitches down. "Hold. Build airspeed. Hold, Rick, hold her down." The end of the runway rushes at me as I hold the turn into the final leg. "Hold her down, Rick. Hold her down. Keep that air speed up to 35 MPH."

At the moment I think we are going to fly right into the earth I bring the stick back. The canard's trailing edge drops. The Eagle rounds out. I am too far to the left, over the scrub grass. This is not my runway. I remember the towing experience. Slight starboard spoileron and the Eagle rolls to the right instantly. We are still too far to the left side of the runway. The runway whips rapidly beneath me. Decisions. Bring her in now on the outside left edge of the runway and land, or up throttle and go around. Snap a glance forward. Enough available runway ahead. Quick right wrist, then to center. The Eagle's starboard wing lowers and the nose centers on the runway. Fine, I say to myself as I push the stick forward. The canard rotates the nose down. Power back all the way to idle. I am on final. My eyes are glued to that white centerline coming at me like the flicking tongue of a white dragon. No time to worry.

Just when I think we'll crash into the tarmac I pull back on the stick and we level off a few feet above the white line. I keep her straight. Hold it. Steady. Compensate for that crosswind. I can feel the Eagle crabbing to the right. Danger! We are oscillating. Quick snap right and left, spoilerons flick up and down, leveling the wings with the horizon. The oscillations stop. Landing is imminent.

A snap glance at the horizon. Wings level. The white center line flashes beneath. Hold on the centerline. Now! Flair. The main landing wheels touch. I feel the slight thunk of contact. Stick forward. Canard all the way up. Nose wheel down. I am back on earth. I flip the kill switch to off, and the Eagle rolls to a stop.

I unfasten my helmet strap, undo the safety harness, disconnect the radio leads, and wait to receive Kris and Bob, who are hurrying towards me, smiles stretching across their faces.

I feel strange. This is my solo. A buzzing permeates my entire body. My ears ring. I fidget with the controls and look at that flat-topped Indian mountain on the edge of the Sandias, and thank my Cherokee ancestors, whose blood still churns within me. I'm glad they could see a brother on his first solo flight. I take a deep breath, lean my head back on the top of the bucket seat, close my eyes, and say a prayer, thankful for my salvation. I am no longer fledgling. I have earned my wings. I am an Eagle.

**STARBOARD SPOILERON** deflection to bring the Eagle XL over the runway centerline. BELOW: Flare just before the main wheels touch down.

☆

# PILOT BRIEFING
# ULTRALIGHT FLIGHT LOG

All records concerning the pilot, engine, airframe and flights must be recorded in the flight log. All situations encountered in flight should be included in the log. All pilot ratings, advancements and insurance claims are dependent on carefully recorded records in the Flight Log. When recording times in the log book, use the following formula. The number .1 (point-one) equals six, 6 minutes. The number 1. (one-point) equals sixty minutes, 60. Example: Log records show the following. Flight Time—1.5. The first 1. (one-point) equals one hour. The point five, .5 equals 5 × 6 = 30. minutes. So the entry recorded flight time (1.5) indicates the pilot flew one and one-half hours on that date of entry.

**LOG RECORDS OF MY TWO FLYING LESSONS TO SOLO.** The log at the top shows the entries made by Chris Williams for my Eagle XL-training. The log below is for the QUICKSILVER MX. Entries were made by Chuck Whittelsey.

# TIME—DATES—NUMBERS

TIME in aviation events such as: departures, navigation, radio communications, log records, estimated time of arrival (ETA) occur within fixed time references and is based on the twenty-four-hour clock system.

Time is referred to as Local Time (LCL), Universal Time (UT) or Greenwich Mean Time (GMT). UT and GMT are also known as ZULU TIME. LOCAL TIME refers to the time of day in the local Time Zone. In the US the local time zones are separated by one hour increments. New York is Eastern Standard Time (EST). Chicago, IL is Central Standard Time (CST). Albuquerque, NM is Mountain Standard Time (MST). San Diego, CA is Pacific Standard Time (PST).

The (AM-PM) 12-hour clock system is rarely used in aviation records. Time varies greatly throughout the world at any given moment. 24-hour time was adopted to simplify this problem of time record consistencies in aviation, weather reports and flight plans.

UT and GMT refer to the Prime (Zero or Zulu—

Z) Meridian, which extends from the North Pole, passes through a town in England called Greenwich, where the Prime Meridian was established in 1884. The Prime Meridian line continues on until it terminates at the South Pole. This ZULU Meridian is the starting point for the 24-hour clock system, that begins at 0000 hrs. and ascends westerly, on hourly (60-minute) increments every 15° longitude. The International Date Line (IDL), is an irregular date boundary line that runs from the poles through the Pacific Ocean and separates the calendar days. On the Asian side it is a day forward. On the Americas side it is a day backward.

UT and GMT are corrected to allow for polar motion and seasonal variations in the rotation of the Earth.

The International Civil Aviation Organization (ICAO) adopted English as the international aviation language for all communications. A phonetic alphabet was also adopted to eliminate confusion between similar sounding letters of the alphabet. Example: Z, E, C, D, G, B, V, and T.

## ICAO INTERNATIONAL CIVIL AVIATION ORGANIZATION PHONETIC ALPHABET

### AIR AND SEA CALL LETTERS

| Letter | Morse | Word | Pronunciation | Letter | Morse | Word | Pronunciation |
|---|---|---|---|---|---|---|---|
| A | · — | Alfa | (*Al*-fah) | N | — · | November | (*No*-vem-ber) |
| B | — · · · | Bravo | (*Brah*-voh) | O | — — — | Oscar | (*Oss*-car) |
| C | — · — · | Charlie | (*Char*-lee) | P | · — — · | Papa | (Pah-*Pah*) |
| D | — · · | Delta | (*Dell*-tah) | Q | — — · — | Quebec | (*Keh*-beck) |
| E | · | Echo | (*Eck*-oh) | R | · — · | Romeo | (*Row*-me-oh) |
| F | · · — · | Foxtrot | (*Foks*-trot) | S | · · · | Sierra | (*See*-airrah) |
| G | — — · | Golf | (*Golf*) | T | — | Tango | (*Tang*-go) |
| H | · · · · | Hotel | (*Hoh*-tel) | U | · · — | Uniform | (*You*-nee-form) |
| I | · · | India | (*In*-dee-ah) | V | · · · — | Victor | (*Vik*-tor) |
| J | · — — — | Juliett | (*Jew*-lee-ett) | W | · — — | Whiskey | (*Wiss*-key) |
| K | — · — | Kilo | (*Key*-loh) | X | — · · — | X ray | (*Ecks*-ray) |
| L | · — · · | Lima | (*Lee*-mah) | Y | — · — — | Yankee | (*Yang*-key) |
| M | — — | Mike | (*Mike*-) | Z | — — · · | Zulu | (*Zoo*-loo) |

**CLEAR ENUNCIATION IS VITAL** in radio communications. Single digit numbers are pronounced as they sound except for the number, 9-nine. Nine is pronounced "NINER" in all radio communications. One explanation is that the word "NINE" in German means "no". Therefore "NINER" was adopted by the ICAO to eliminate any confusion.

Two digit numbers or more are enunciated as single-digit numbers. Example: 1985 is spoken—ONE NINER EIGHT FIVE. 2,001—TWO THOUSAND ONE. 700—SEVEN HUNDRED. 990—NINER NINER ZERO.

### NUMBERS IN RADIO COMMUNICATIONS

1-ONE — 2-TWO — 3-THREE — 4-FOUR — 5-FIVE — 6-SIX — 7-SEVEN — 8-EIGHT — 9-NINER — 10-ONE ZERO — 15-ONE FIVE

23.9-TWO, THREE, POINT, NINER (THE USE OF 'POINT' IS USED WHEN REPORTING RADIO FREQUENCIES)

24.09-TWO, FOUR, ZERO, NINER (FOR ALTIMETER RADIO REPORTS THE DECIMAL POINT IS OMITTED IN COMMUNICATIONS)

1,538-ONE THOUSAND, FIVE, THREE, EIGHT (FOR CEILING ALTITUDES, FLIGHT AND WIND LEVELS)

11,000-ONE, ONE, THOUSAND

# FLIGHT TRAINING

**QUICKSILVER® MX** 'THE ORIGINAL'

| | |
|---|---|
| Date: | Thursday 29 September 1983 0700 hours |
| Weather: | Clear |
| Temperature: | 68° |
| Wind: | 0-4 Knots Westerly |
| Place: | Lake Elsinore Flight Park CA |
| FAA-Certified Flight Instructor: | Chuck Whittlesly II |
| Aircraft: | Quicksilver MX II |
| Training: | Dual-Preflight-Taxi-Flight Controls |
| Student: | Rick Carrier |

## OPERATIONS/EQUIPMENT

### SPECIFICATIONS

| | |
|---|---|
| Powerplant | 1 Rotax Model 377 Twin Cylinder, Two Cycle 368.3 cc, 33.5 hp @ 6500 rpm |
| Recommended TBP | 250 hr |
| Propeller | 52 in x 32 in |
| Length | 18 ft 1 in |
| Height | 9 ft 8 in |
| Wingspan | 32 ft 0 in |
| Wing area | 160 sq ft |
| Wing loading | 2.28 lb/sq ft |
| Power loading | 15.6 lb/hp |
| Seats | 1 |
| Minimum flight crew | 1 |
| Empty weight | 239 |
| Useful load | 290 lb |
| Payload w/full fuel | 260 lb |
| Max takeoff weight | 525 lb |
| Fuel capacity | 5 U.S. gal |

### PERFORMANCE

| | |
|---|---|
| Takeoff distance, ground roll | 69 ft |
| Takeoff distance, 50 ft obstruction | 220 ft |
| Rate of climb | 800 ft/min |
| Max level speed, sea level | 52 mph |
| Max operating altitude | 10,000 ft |
| Landing distance, 50 ft obstruction | 150 ft |
| Landing distance, ground roll | 60 ft |

### CRUISE PERFORMANCE CHART
(Speed/Range at sea level)

| | |
|---|---|
| @ 55% power (18.43 hp) | 38 mph/95 mi |
| @ 65% power (21.78 hp) | 41 mph/66.1 mi |
| @ 75% (25.13 hp) | 46 mph/52.3 mi |

### FUEL | FLOW

| | |
|---|---|
| @ 55% power | 2.0 gph |
| @ 65% power | 3.1 gph |
| @ 75% power | 4.4 gph |

### LIMITING AND RECOMMENDED SPEEDS

| | |
|---|---|
| Vx (Best angle of climb) | 35 mph |
| Vy (Best rate of climb) | 37 mph |
| Va (Design maneuvering) | 50 mph |
| Vne (Never exceed) | 63 mph |
| Vs1 (Stall, power off) | 24 mph |
| Landing approach speed (1.3 Vso steady flight speed while landing) | 31 mph |

All specifications are based on manufacturer's calculations. All performance figures are based on standard day, standard atmosphere, at sea level and 175 lb. pilot weight, unless otherwise noted. Operations/equipment category reflects this aircraft's maximum potential.

**DAY 1**

I left Albuquerque with respect and admiration for the Eagle XL and its builders, American Aerolights Company. They had provided me with the opportunity to learn to fly. Forever I will be grateful to them for giving me the chance. A dream fulfilled.

As Bob Milliken shook my hand at the airport on Sunday morning following my solo, I knew that we would someday fly together. As I walked towards the waiting jet to San Diego, I thought of Kris Williams, his wide smile and friendly greetings. I had enjoyed them every morning throughout the past week and, today, I was heading to a new adventure. I felt good, indeed.

Now that I had soloed, buckling up in the starboard window seat of a commercial jet and having a friendly, blond, world-traveling San Diego socialite sitting next to me as a companion, brought back the exhilarating spirit of flight. I could feel the blood pump the moment we started the takeoff roll. My mind started to click. Counting seconds. Thousand-one. Thousand-two. The mains of the 707 broke from the runway surface in liftoff rotation in exactly 21 seconds. I watched the Earth crawl away beneath as we did the climb out of the Albuquerque TCA (Terminal Control Area).

My eyes focused sharply on the details down on the surface of the deck as we gained altitude. Observing the land from the open seat of an Ultralight on takeoff and cruise conditioned my thinking. I was looking for landing sites. Strangely, my eyes picked out details on the desert surface I would never have noticed before.

Glittering sunlight reflected off the surface of water in storage tanks and ponds. Discolorations between damp and parched land were distinct. Winding dirt roads that seemed to disappear into the dry desert and then re-appear on the far side of the mesa, got lost again in a valley and then re-appeared to track straight and true on level ground. Tiny shacks and brown buildings were situated in the middle of nowhere. I wondered who lived in such isolation.

We flew over a military airfield where B-52 bombers and other warplanes could be clearly seen lined up in their parking berths within the MOA (Military Operations Area). By tracking my eyes along the fence and access road that snaked around the base, I visualized the blue hash marks on the sectional aeronautical chart that outlined the restricted MOA zone.

119

The endless panorama of the desert and the objects down there held my attention until the jet climbed up through the clouds and into sparkling sunlight and leveled off at 32,000 ft. AGL. The cabin attendants served food and drink and I had a wonderful time with my seat mate. The flight to San Diego was all too short.

I picked up my rental car, drove my socialite seat mate to her home that overlooked the Pacific, made a date to meet again in New York and drove a few miles north on the interstate to the Marine base at Camp Pendelton where I would be the house guest of old friends; Major Edward Whitehead, USMC, his wife Bonnie and two lovely, delightful daughters, Mitzi and Christi. I had a fabulous time with the Whiteheads and their friends. They were the most gracious host and hostess I have known. Eddie and I complemented each other in spirit and adventure. He had, like myself, always had a desire to fly. Being around me during my flight training flamed his interest. Before I left California I would see to it that Eddie achieved his wish. He did.

Being on a Marine base and seeing the boot trainees every morning out on their PT (physical training) exercises as I drove by on my way to training, had a conditioning effect on me. I developed the "Boot" attitude rather than the "Media guy from New York" who was writing a book. I approached flight training from the grass roots. I insisted on taking the same course as anyone who walked into a dealer, planked his money down and said "I want to learn to fly one of your airplanes." It was how they taught him or her that I wanted to write about, with no special favors. Having flown the Eagle XL gave me an edge but I wanted no short cuts taken.

The thirty-eight mile drive from the Marine Base to the Eipper Aircraft Headquarters at Rancho California near Temecula was a travelog. Past orange groves... rocky mountains... eagles circling high overhead... bluffs overlooking Lake Elsinore... winding, twisting roads that snaked steeply down mountainsides to the valley floor then to the home of the Quicksilver MX. The factory and home office of Eipper reminded me of the spacious pharmaceutical complexes located on rural settings in Connecticut and New Jersey. The sprawling Eipper plant sparkled with cleanliness and manicured exterior maintenance. Everything seemed in place and as my visit there lengthened, I discovered my first observation to be true. Lyle Byrum, the president of Eipper obviously ran an organized company. His leadership was evident.

I drove into the parking area in front of the ranch-style building and found a spot under a flowering trellis. The handsome blond Eipper Aircraft receptionist greeted me with her brightest Southern California smile and hospitality. I sat down in a comfortable lounge in the tastefully appointed Spanish ranch style reception area. I sipped hot coffee and talked with a propeller manufacturer who was also waiting for his appointment with Lucky Campbell, Eipper's V.P. Sales, who had arranged for my visit and training.

"Hey, Rick!" I looked up. There was Bruce Noll, leaning over the edge of the balcony. His smiling face and waving hand beckoned me up. I slung my cameras and recorders over my shoulder and hurried along. Bruce Noll, Eipper's Special Projects Coordinator, and I re-lived the fun and excitement we had together at Oshkosh. I showed Bruce the pictures I'd taken of Walter Kole and his hilarious antics as he cast his fishing lure from the seat of the float-equipped Quicksilver. The Quicksilver was sitting in the middle of a lily pad on the edge of Lake Winnebago. Walter had a tiny bass on the hook and we all said it should be his dinner that night at the barbecue and dance Eipper was hosting at the farm behind the flightline at Whitman field. It was a fun party and Walter was the only one who made it under the limbo bar set very low. The party was for all the people in the ultralight community. They came from all over the world. The talk was ultralights, exclusively. The party ended quite late on that warm August

**EIPPER AIRCRAFT INC. – TEMECULA, CA.** Bruce Noll, Eipper's Special Projects Coordinator, here standing in front of the main entrance of Eipper's headquarters and factory. He saw to my every need while I was taking flight training on the Quicksilver MX.

Wisconsin evening with a fireworks display. Eipper Aircraft has style. Here I was, in California to learn to fly the Quicksilver MX from the people who build them.

Bruce took me on a guided tour of their headquarters. I re-kindled friendships made at Oshkosh. Lyle Byrum, the president, greeted me warmly and hoped I had good weather for my flying lessons. Then to Lucky Campbell, the Vice President of marketing and sales. We had a long conversation on the subject of ultralight safety as he pointed out the quality controls in building this ultralight. I was impressed.

Bruce introduced me to Charles B. (Chuck) Whittelsey, Eipper's Director of Flight Operations, my flight instructor. I liked my 26-year-old certified flight instructor immediately. He outlined my training schedule and then we went to see officers from the Downey California Policy Department and their Quicksilver patrol plane.

1. Rib
2. Trailing Edge Spar
3. Leading Edge Spar
4. Compression Strut
5. Kingpost
6. Spoiler
7. Wing Cover
8. Rudder Leading Edge
9. Rudder Frame
10. Rudder Compression Strut
11. Elevator Frame
12. Stabilizer
13. Tail Skid
14. Rudder Brace
15. Tail Brace
16. Tail Mount
17. Push/Pull Tube
18. Teleflex Cable
19. Tail Boom
20. Propeller
21. Reduction Unit
22. Root Tube
23. Landing Gear Down Tube
24. Seat Support Down Tube
25. Axle
26. Main Wheel with Pant Covering
27. Axle Strut
28. Seat Mount Assembly
29. Tension Strut
30. Nose Strut
31. Nose Wheel
32. Foot Bar Assembly
33. Control Stick
34. Triangle Bar Down Tube
35. Engine
36. Fuel Tank
37. Diagonal Strut

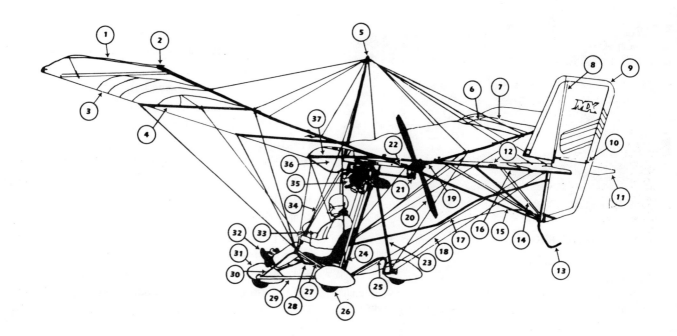

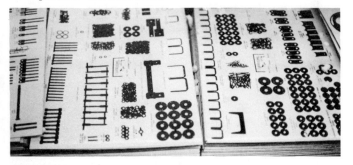

**QUICKSILVER MX PACKAGING** above and below, indicate the pain and effort Eipper Aircraft employs to assure quality control and simplicity of assembly. Each part is sealed and numbered in a plastic covered container.

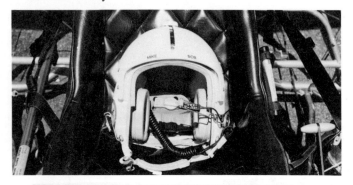

**QUICKSILVER MX ULTRALIGHTS** assembled at the Eipper Aircraft factory.

**HELMET PLACED ON THE PILOT'S SEAT.** The first step in the pre-flight procedure.

**WING LEADING EDGE.** During pre-flight you must run your hand along the entire length of the leading edge of the wing, looking for dents or bends in the tubing under the wing fabric. If irregularities are discovered, the fabric must be removed and the wing tubing examined. If bends or dents are found they must be corrected before flight. The ultralight is grounded until repaired.

# PREFLIGHT QUICKSILVER MX

You discover many things when you preflight an airplane. During this first contact with the bird you are about to fly you notice the materials and the way it was built.

Your life depends on the airplane staying together in the air as you pay very close attention to everything about her as you feel each item as you examine it.

I stood back and looked at the Quicksilver MX. I immediately liked the rake of her wings dihedral and the way she sat on her tail with the nosewheel pointed jauntily into the air as though she was about to leap into the sky like a grasshopper. She has spunk, I thought. I was enjoying this introduction very much.

I cleared my head of distracting thoughts as I walked up to the MX and began my first preflight. I knelt as I placed my helmet on the seat and swung the stick side to side, while watching the rudder swing back and forth. The action was smooth and silky. I slid my left hand along the teleflex cable looking for flaws, kinks, loose connections or anything unusual as my right hand rotated the stick fore and aft as my eyes fixed on the elevators fluttering up and down. This control also was smooth.

After each item was examined, I double-checked the preflight card and repeated aloud the instruction and then did it to the letter. I placed both hands on the foot pedals and moved them fore and aft and closely watched the shadows of the spoilers on the upper surface of the wing move up and down. The spoiler control cables were smooth and un-encumbered as were the pulleys. All worked well.

I raised the plastic seat cushions and examined the main frame attach points and saw that the nuts were seated so that two threads were exposed above the nut. Then I examined all the struts around the seat. Looking for kinks and dents in the tubing. OK.

I spun the nose wheel and closely examined the axle. I squeezed the tire. Pressure OK. All of the fittings for security safety pins and locknuts were in place.

Now a look at the entire landing gear. I noticed the white left main wheel pant was cracked at the attachment point and it had a slight wobble. I brought this to Chuck's attention and after examining it he said: "The pant is fine. Is is not a hazard to flight nor will the crack interfere with the main gear function." I went on with my preflight. Next came the nose struts. All in fine shape. The foot bar and tension struts and all of the connecting hardware. Every item, I studied and examined very carefully before continuing on.

On to the next item; the triangle bar tubes and the clusters of flying wires that are attached, and their fittings. I took my time on this one for, of all the accident reports I have read, this is where many of them began. A kinked or frayed wire, breaks or gives way under the strain of flight and the wing collapses and down she comes. All was ship-shape. I went on.

I straightened up and looked down along the port right wings leading edge. Straight as an arrow. Next I slid my hands along the edge as I walked towards the wing tip. Massaging the fabric as I walked, searching for a dent in the fabric-covered metal wing tube. In addition, the wing fabric came under close scrutiny. Here I was looking for tears, abrasions and the general condition of the covering. I tilted my head under the wing and looked up into the corners, looking for anything that did not look quite right. All OK.

The leading edge wires, spoiler control arm, the bungee elastic return and the spoiler control line were all in smooth operating condition.

Slowly I walked around the port wingtip. Touching and feeling everything along the way. I stopped and grasped the compression strut and rocked the MX up and down. I was looking for any give or play in the wing. It was solid as a rock. All ribs securely in place.

Walking along the rear of the port wing I eyeballed the rear trailing wires and their attachments to the king post. Good.

I knelt and reexamined, from the rear, the main wheels of the landing gear, axles and hardware and the down-tube condition. Excellent.

Standing up I directed my focus on the engine and all of its components, starting with the reduction drive unit. The belts were clean and healthy looking. The tension of the belts, the right deflection measure and then sliding my hands over the warm engine, looking for anything loose or out of place on or near the reduction drive belts or flanges. All clear.

The tail rib booms were securely in place and the

**CRITICAL JUNCTION** of the flying wires where they cluster on the the triangle bar. Any flaws or failure here and the wing can and possibly will collapse when under flying stress.

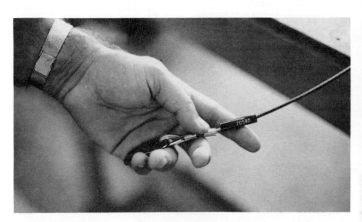

**ALL FITTINGS** must be free of kinks, bends or abrasion. Any signs of wear or fraying and the entire assembly must be replaced.

**SAFETY RINGS** must be on all junctions where specified. A missing safety ring could and in many cases will spell disaster when flying.

**THE ENGINE** and all components must be carefully checked for integrity. A failure here and a forced landing is immediate.

attachments to the main frame and the Pip pins on the tail mount were secured and locked in place.

Before I touched the propeller I made certain the Kill Switch was OFF. Now the propeller. I slid my palm over the smooth varnished (Epoxy) surface and looked closely at the wood grain and surface. I paid particular attention to the tips. There were no nicks or abrasions. The propeller hub and nuts were tight and the lock bolt through the hub secure.

The clearance from the propeller tip and the elevator push-pull tube were correct. I had heard of situations where the prop came in contact with this tube and nearly caused a serious accident. This was a critical measure as was prop clearance from the tail booms.

The expansion chamber (muffler) bracket, propeller shaft and coupler were secure.

Both tail brace tubes were attached to the 1" channels. Pip pins in place and secure.

Tail booms attached and Pip pins secure.

I followed the teleflex cable from its attach point on the rudder horn. A safety pin and a clevis pin held the teleflex fork in position on the rudder horn. I slid my hands back over the rudder brace attach point on the outboard side, down along the lower tail wire, (secured with Velcro straps), inside the surface of the landing gear down tube, then to the outside of the gear strut and under the nose strut and up to the control stick where the teleflex cable was secured with a safety ring and pin.

Standing I directed my attention to the kingpost and all of the hardware attachments there. I tugged at each of the Kingpost wires and studied the attachments before giving it my OK.

The elevator tube was free of nicks and bends but lo-and-behold when I was checking the elevator hinge assembly for security, the starboard hinge attachment was missing a safety pin. I mentioned this to Chuck and he produced the pin from his pocket and handed it to me. I replaced it and went on. My heart quickened. He was testing me on my first preflight. I narrowed my eyes as I went on to the next item on the preflight card.

The rudder and elevator attach points were all secure, as was the restraining line from the push-pull elevator tube to the main gear axle to ensure prop clearance when the plane bounced during take off, Taxi or hard landings. All OK.

**THE IGNITION SWITCH** must be in the OFF position during the pre-flight inspection. If for any reason the ignition switch is ON, a serious accident can occur when the pull starter is checked or when examining the propeller.

**EXPANSION CHAMBER** (muffler) bracket. Should one or both bolt positions be loose or missing, the muffler could fall down and cause the engine to malfunction. In flight this is disastrous.

**RUDDER HORN SAFETY RING AND PIP PIN.** Both must be secure and in place. If missing, the rudder will fail and the ultralight will cease to fly under the pilot's control. A crash is imminent.

**ELEVATOR HINGE ASSEMBLY.** The safety ring holds the elevator pin in place so the elevator can freely move up and down. If the safety ring is missing, the vibrations of flight can and will dislodge the pin. With the pin missing, the elevator will violently flap up and down and quickly tear off the tail. Immediately the ultralight will pitch nose down and descend vertically to impact.

Spoilers lay flat against the wing surface. Perfect.

I moved down the starboard trailing edge of the wing, feeling the spar and checking all the attached wing wires for security. Fine.

All ribs in place and secure. I grasped the compression strut and rocked the starboard wing. Solid.

I moved around the starboard wingtip looking for problems. My hands work along the leading edge spar, again looking for dents, abrasions or tears in the wing fabric. All OK, including wing wires and spoiler control system.

Root tube bolts secure. Kingpost, height and attachments, OK.

Full tank, 3/4 full. Mountings, secure. Fuel valve, On.

Fuel line, secure. No kinks. Clamps on fuel tubes tight. Fuel filter and tank crossover, clear no impurities. I asked Chuck if we should drain it anyway and he said he had done it before flying out to Lake Elsinor. I ran my fingers over the entire engine looking for any signs of leaking gasoline. None.

Carefully I surveyed the engine, pull starter, spark plugs and ignition wire attachments. I had heard tell of spark plug wires popping off on take off. Nasty business. These plug wires were tight down. I looked at gasket seals for signs of wetness or oil stains. None.

I had completed the preflight and one of the things that I noticed about the Quicksilver MX was the care and pride in workmanship that Eipper Aircraft had taken in building her. No rough hacksaw marks here as I had seen on other Ultralights at Oshkosh. All tubes had polished or machined ends and capped. The tubes themselves were anodized an attractive light blue. All fittings were Aircraft "AN" quality. The sewing on the wing and tail surfaces were neat and wrinkle free. Everything about the MX was first rate. I was looking forward to flying her. I turned to Chuck.

"I've finished the preflight. The only thing I found was the missing pin on the elevator assembly but, to be sure, did I miss anything else?"

Chuck smiled. "NO. But you will not fly the MX today. I will train you in the MX II, the two-place trainer. I have preflighted the trainer, so put on your helmet, hop in and buckle up. You are about to have your first lesson."

I finished taking pictures of the Quicksilver MX after my preflight when Chuck walked up to me and said: "Here's your helmet. Put it on, get in and buckle up."

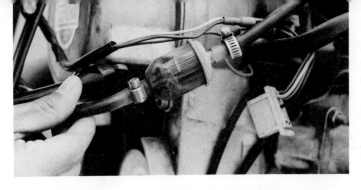

**THE FUEL LINE** is a critical pre-flight inspection point. Contamination of any sort or if the Fuel Valve is *closed* means the engine may start and run long enough for a take off then stop in the middle of the climb out. Crashes, forced landings and aborted flights have been caused when the pilot neglected to pre-flight the Fuel Line.

**THE PULL STARTER** must never be tested when the ignition switch is in the "ON" position. Before pre-flighting the Pull Starter always double check to make certain the kill switch (ignition) is "OFF". When pulled, the Pull Starter should turn the engine over then retract smartly back up into the Pull Starter Housing. If the Pull Starter Line sticks or does not return, do not fly or start the engine until the Pull Starter is fixed. The ultralight is grounded.

**QUICKSILVER TWO-PLACE** training model. The instructor sits in the right hand seat and the student in the left. There are two sets of controls. One for the instructor and one for the student. Chuck is shown here as he is about to attach the Parachute Carbiner to the cable that is secured to the Root Tube of the Quicksilver.

**FUEL LEVER IS IN THE DOWN-ON** position before turning the Ignition Switch to the ON position.

**IGNITION SWITCH IS IN THE ON** position. To either side of the Ignition Switch is the KILL position, Fore and Aft.

**THE THROTTLE IS BROUGHT BACK TO THE IDLE** position before starting the engine.

**CHUCK IS GOING THROUGH THE FINAL STAGES** of the control checks before starting the engine. The entire area around the ultralight is checked for any activity that could interfere with a smooth Start, Run-up, Taxi to the Active Runway and Takeoff.

The "D" ring closure on the helmet's chin strap gave me trouble. I was all thumbs. Chuck's fingers threaded the strap through the rings and tugged the clasp up tight. I raised my chin in recognition. I adjusted it and settled myself into the left-hand seat, slid into the shoulder harness, and buckled up. The headset crackled as Chuck plugged in the jack and said: "Are you reading me?"

I nodded affirmative. Chuck went on. "Watch what I do and remember the sequence."

I settled into my seat and watched.

Chuck hooked his parachute carbiner to the aircraft recovery system cable attached to the root tube and checked the flight controls. I closely watched as he manipulated every one. Stick port and starboard. He worked the rudder left and right. Stick fore and aft. He worked the elevator up and down. His left and right foot alternately depressed and released worked the spoilerons up and down. He reached over his head and made certain the fuel lever was pointed down. Then he brought his hand down beside the throttle and flipped the ignition switch to the forward "ON" position, reached back and depressed the choke to the "DOWN" position, set the Throttle to the rear "IDLE" position, took hold of the starter handle and shouted: "CLEAR PROP" and pulled down hard. The Rotax engine roared to life. Chuck let the engine run for a few moments then shut off the Choke.

The taxiway and runways at Lake Elsinore were dirt surfaces. The prop blast whipped up clouds of dust into a swirling storm behind us. Chuck spoke. "Always look behind before you start the engine so you will not blow dust and stones into anyone or parked aircraft. You steer the Quicksilver with the rudder. The nosewheel is fixed straight ahead. To turn, you advance the throttle to full to achieve breakaway thrust. Once the Quicksilver starts to move forward, swing the stick towards the direction you want to go. The rudder swings out into the prop blast and the force of the air against the rudder forces the tail around and points the nose in the direction you want to go.

Back pressure on the stick, raises the elevator into the prop blast and takes pressure off the front nose wheel which assists in making turns, particularly in grass or bumpy surfaces."

Chuck smoothly taxied the MX to the head of the runway and lined the nose straight down the center.

Chuck spoke: "The last final check before takeoff is called The *Before Takeoff Check.* Rework all of the controls. Look for any stiffness or bindings as you work them back and forth." He put my hand on the controls and I worked the stick back and forth and right and left. I turned and looked at the rudder and elevator flutter up and down. "Next thing you check is the engine. Fuel valve down. Choke up. Throttle slowly forward to a full runup. You keep from going forward by using foot pressure on the brake post over here to my right. Bring the throttle back as you look around for any planes taking off or landing. Look at the wind sock to see what the wind is doing and how hard the wind is blowing. Wind speed at this stage of your training should not exceed 5 knots. Relax your body and bring your toes back on the foot pedals to release pressure on the spoilers so they are flat against the wing. If they were up the wing would not fly. The next step in the takeoff procedure is to advance the throttle slowly forward."

The Quicksilver started to roll and as the speed increased Chuck slid the throttle all the way in. The Rotax roared. My grip on the cage down tube tightened. The nose lifted and pitched up steeply. We broke away from the earth in seconds and climbed steadily upwards. At about thirty feet AGL he leveled out and increased his airspeed by holding level flight instead of climbing. Over the end of the runway he started to climb again and we remained in that climb until we rounded out at one thousand feet AGL.

The view of Lake Elsinore and surrounding hills was spectacular. We were following the southern rim of the Lake. Below us was a parachute training field. A DC-2 sat beside the main hangar. She looked trim from this altitude. My thoughts were interrupted when Chuck called my attention to his left index finger. He had his fingertip touching the top of the stick. He was flying the plane with his fingertip. He motioned for me to take hold of the stick. I did and started to fly the MX. She responded smoothly. A little pressure to rear and up she goes. I rotated the control to the right. The MX gently tilted her wing to the right. I brought the stick back to neutral and the wings leveled with the horizon. It felt great.

"Make all moves smoothly, Rick. Fly it a while. Get the feel of it. I'll be here."

**CLEAR PROP** is always shouted before pulling down on the Pull Starter. The engine should never be started unless the ultralight is in the START - STOP location designated by the Ultralight Flight Park Rules. A final check of the area all around the ultralight must be done before the engine start. The Prop Blast can be severe and anyone or other aircraft behind the ultralight can be seriously damaged if too close to the rear.

**TAXI TO THE ACTIVE RUNWAY** must be slow and easy. There is absolutely no room for any playing around or a fast taxi to, or from, the active runway. The rudder is used for turning on the Quicksilver MX.

**FINAL CONTROL AND AREA CHECK** before takeoff. Many pilots neglect this very important procedure and as a result of their neglect, have had serious accidents. This Final Check is a vital Pre-Take-Off procedure.

**LIFTOFF** occurs in only a few seconds once full throttle is applied. As soon as Lift-Off Rotation has been achieved, Chuck leveled off over the runway to gain air speed before attempting to gain altitude. Using this procedure assures proper air speed (35mph) for the climb-out. Pilots that neglect this procedure have had serious trouble when the engine quit or if some other malfunction occurs. Remember. Air Speed must be achieved and maintained during takeoff.

I took the stick and got acquainted with the Dual MX. She was a fine ultralight. The trim was exceptional. A bank and I released stick pressure. She righted herself. The Quicksilver's dihedral of the wing (positive stability) was responsible for the quick return to level flight.

I noticed our feet. Chuck's to the right and mine to the left. When I rotated the stick forward the nose went down. So did our toes. Back on the stick and they went up over the horizon. The degree of elevation of our toes over the horizon indicated the angle of climb.

I kept slight back pressure on the stick. Our toes went into the sky. Chuck wrapped his fingers around mine as he spoke. "I'm going to take it up into a stall." He pulled the stick all the way back. The wings started to buffet and shake. Suddenly the nose dropped and the buffeting stopped. It was a smooth recovery. Chuck gave the stick back to me, saying: "You try it."

It required a lot of back stick pressure to hold the MX into a stall. When she fell out and started down the controls returned and I leveled out.

Chuck asked me to make a right hand turn but to hold it level. I tilted the stick to the left. Chuck increased the throttle as the MX started into the turn. The nose dipped. Chuck laid his hand over mine and gave slight back pressure but held the stick in the turn position. The MX leveled out as she held into the 10° turn. With his hand still over mine he rotated the stick to slightly beyond center and at the same time came back in a smooth swinging "U" then back to center. The MX was flying straight and level.

"Now you try it, Rick."

Within fifteen minutes I had the turn, while keeping it level, under control. I did left and right hand turns up to 20° bank.

We were still about 1,000 AGL and on the western side of the lake opposite the ultralight air park. Chuck said: "Turn and fly around the edge of the lake. We'll do some practice landing approaches."

I did a shallow 180° turn and as I came to my heading I lowered the nose in a gradual descent.

Chuck said: "Stay over the shoreline. As you descend keep a sharp lookout for other airplanes. As you near an airport assume that there will be airplanes about. Do not depend on other pilots seeing you. You see them first. See and avoid. Also, keep a close look for alternate landing sites in case the engine fails. Notice the wind direction. The wind may have changed since we took off. Always check the wind."

I scanned the entire area. An Eagle XL was just taking off. It was exciting to see the ultralight I did my solo in coming up to join us. Chuck continued.

"When you come in on the airpark traffic pattern enter the downwind leg at a 45 degree angle at an altitude of 500 to 300 feet. If an ultralight is in front of you in the downwind leg, they have the right-of-way. Give them plenty of room. Then fly level along the downwind leg until you are opposite the point on the runway where you want to touch down. At that point you begin a gradual descent."

Chuck had the controls in his hand now.

"Watch and make mental notes of what I am doing and saying. Notice that we are almost half-way down the runway and we have lowered our altitude to about 200 to 300 feet. We continue on down, gradually until we reach the end of the runway. As soon as you pass the end of the runway you turn into the Base Leg. The altitude is now about 100 feet. Pick out the touchdown spot on the center of the runway, then bank and turn as you lower the nose. Keep plenty of airspeed, and point the nose right at your touchdown spot."

Chuck banked the MX and we seemed to head straight down towards the runway. Actually we were descending at about a 45° angle, but to me it seemed very steep indeed. Chuck went on.

"When you think you are going to hit the ground, round out and fly level over the runway and just when you think you are going to set the wheels down, pull gently back on the stick, and the main wheels will touch first, like this." The mains touched then the nose wheel. We rolled along the runway a short distance and Chuck ran up the throttle and we took off again.

We flew touch and goes until I had it together. We had been in the air almost an hour. I was making a descent on the downwind leg when I noticed an ultralight flying up on my starboard wing. It looked like he was going to head me off.

I asked: "That ultralight moving in on my right wing. Should I go around and let them through." Chuck looked at the approaching ultralight. "You have the right-of-way. You are in the downwind leg. But remember, ultralight pilots do not necessarily know all the rules. It is better to go around than take a chance. Remember you are the pilot in command. You make the decision."

**CLIMB-OUT.** Once the air speed reaches 35mph the climb-out procedure can be initiated, never before and then turns can be made.

**AIR TRAFFIC.** You must always scan the sky for other aircraft. It is the responsibility of the Pilot in Command to see and respond to encroaching aircraft, and to stay a safe distance away from them. Each ultralight has what is known as a dead sight line zone where the pilot cannot see by direct scan. This spot on the Quicksilver is above and behind the wing. By constant scanning of the sky, the pilot can adjust for this dead zone and know what is above and behind as well as in front, to the sides, and below.

**GLIDE PATH** is straight and in line with the center of the runway. Only minor control inputs are required during the final descent. The power remains constant during this final approach.

**FLARE.** With continued back pressure on the stick, the ultralight pitches the nose up in a Flare and the rear main wheels touch the runway surface.

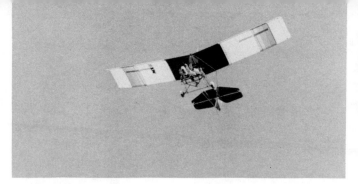

**ALTITUDE** — 1000ft AGL is the training altitude at Lake Elsinore. Each flying lesson lasted about one hour. In that time the student has an opportunity to experience and try: Turns, Banks, Stall, Practice aerial landing approaches, descents, roundouts and flares with the safety of plenty of altitude.

**APPROACH DESCENT.** The air speed and descent is constant. The pilot's vision is focused on the point where the touchdown is to be made.

**ROUND-OUT** decreases the glide angle by rotating the control stick back and raising the elevators. This maneuver is executed about ten feet AGL and is maintained until the ultralight descends to about two feet AGL. Airspeed has decreased to just above stall.

**ROLL-OUT.** Care must be taken here to assure a straight course down the center of the runway. Positive control must be used throughout the roll-out until airspeed is bled off.

TAKEOFF

CRUISE

CLIMB-OUT

APPROACH DESCENT IN THE FINAL LEG.

# QUICKSILVER MX

**DAY 2**

The excitement of the first days training had me pumped up tight. I was ready to go for the second. I arrived at the airpark at Lake Elsinore just in time to get a series of good pictures of Chuck flying in from the Eipper Aircraft factory at Temecula several miles to the South of the lake.

During my morning ritual preflight I discovered that Chuck has set the fuel lever to the "OFF" position. I corrected it and turned to Chuck and said: "She is ready for flight."

He handed me my helmet as he said: "Get in and buckle up."

I carefully watched Chuck as he went through the paces of his pre-takeoff check list. He was pro all the way. He manipulated the flight controls with gentle firmness. First the rudder then the elevator and spoilers, then his fingers knowingly checked the fuel valve, making certain it is "DOWN" in the "ON" position. His right foot pressed down on the brake post as he flipped the ignition switch to "ON" then reached overhead, grasped the starter handle and shouted loudly; "CLEAR PROP", and yanked down hard. The Rotax flamed to life. Clouds of blue smoke and dust churned behind us as he slowly advanced the throttle for the full power run-up. Looking at him I could tell he was listening for engine misses or roughness in the ignition system of the engine.

All the while his eyes scanned everything from the area behind us to the sides in front and above. Nothing escaped his attention to detail. He brought the throttle back to Idle and let the engine crank for a moment then released the brake and slowly began to advance the throttle.

We started to move forward on the takeoff roll. I counted seconds. Two seconds. Three seconds. My heartbeat quickens as it always does on takeoff. Four seconds. Five and the stick comes back. We rotate into the air as the Rotax winds up tight. We gain altitude. Swiftly the runway retreats behind us. The thrill is there for me. I marvel at the climb-out smoothness. A quiet smile etches my cheeks. This is the experience of flight. I love it and from the look on Chuck's face, so does he.

We rounded out and leveled off at 1,000 ft. AGL. Chuck turned the control stick over to me, saying: "We'll work on turns, banks and air traffic patterns today. Fly over to the west side of the lake and make a 45° turn North."

My mind clicked into an intense concentration mode. "I must remain loose," I thought. "No stiffness. Relax your fingers on the stick, Rick." The Quicksilver flew steadily along all by herself. I had my fingers wrapped around the stick but not touching the sponge rubber grip. I was amazed at how stable the Quicksilver was and how smoothly she responded to even the slightest control input. A gentle touch, right or left and the wings acknowledged with a smart short dip then back to straight and level flight.

Slowly, I applied forward stick pressure. The nose pitched into a gentle descent. I held it there. Feeling, testing just how much pressure was required to hold the ultralight in the shallow dive. It was not very much. I calculated about two pounds of forward pressure. I rotated the stick back into a climb attitude and held it there until we were back to cruise altitude, 1,000 feet AGL, where I brought her back to level flight.

As we approached the South Western corner of Lake Elsinore below, I rotated the stick to the starboard and brought the Quick into 20° righthand bank while holding the nose up to keep her level.

Chuck spoke. "You have a throttle control on your side. If you need more thrust to hold altitude in the turn use it."

My left hand gripped the throttle handle and inched it slowly forward. I was cautious. This was the first time I'd used the throttle on the Quicksilver. It responded smoothly. I advanced the throttle just enough to keep the nose level with the horizon while still in the bank. When the Nose pointed North I brought the stick back to center but the Quick continued her roll. Chuck took the stick.

"Watch me, Rick, and I'll show you how to make the turn and roundout directly on the heading without oscillating like you had the plane doing just now."

I intently watched as he made the maneuver. He banked hard to port. Almost 30° then with a smooth wing of the stick back in a curve continued on around without stopping, past the central neutral position and then back to center. The Quick came out of the bank and did the roundout directly on course without a wobble to the wings. He handed me the stick with a gesture. I took it and repeated the maneuver. Within three or four tries I had the technique together.

I began to feel and understand the meaning of the

# PILOT BRIEFING
## LANDING PROCEDURES

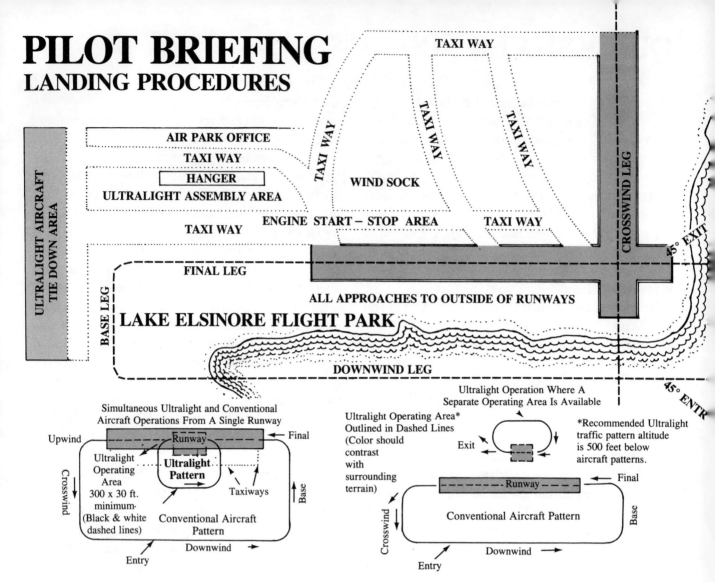

## RUNWAY PERSPECTIVE

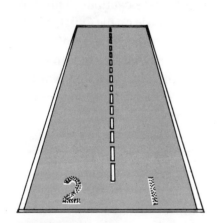

**PERSPECTIVE CORRECT** for a proper glide slop on the final approach to landing. Be alert for cross wind shifts and other ultralights landing or taking off.

**GLIDE SLOPE TOO HIGH** — Perspective of runway indicate the approach is too high. Abort landing and go around and re-enter the landing pattern. CAUTION — Always scan the entire landing area and pattern for other aircraft entering, leaving, taxing or flying around the patch. Mid-air collisions at this point can and do happen all to frequently because pilots did not scan the area thoroughly before making abrupt alterations in their flight pattern.

**GLIDE SLOPE TOO LOW** — Perspective here indicates that the glide slope is too low for a landing without increasing thrust to gain altitude. A typical result of a too low approach is; The pilot increases thrust. The ultralight gains altitude. The pilot decreases thrust. The ultralight sinks as it loses altitude. Problem: The pilot has flown way past the chosen landing spot and risks flying right off the runway and landing in trees, lakes, buildings or worse. It is the wise pilot that aborts the landing, and makes a go-around to re-enter the landing pattern.

**THE PERSPECTIVE OF THE RUNWAY** during final approach should always remain constant as shown in illustration 1 above. Up and down "porpoising" indicate pilot over control and if not corrected before touchdown, a hard landing or worse, an undershoot or overshoot of the runway is predicted. The over-undershoot problem is usually corrected by gentle but positive handling of the controls. Remember: It is up to you, the pilot in command, to control overcontrol.

# QUICKSILVER MX

light and gentle touch. The Quick flew without hard control. As a matter of fact the gentler the control input, the smoother she flew. The secret I was beginning to understand was being "SMOOTH" while maintaining positive, yet firm command of the controls.

Chuck was a hard but excellent flight instructor. He allowed plenty of room for the student to discover the flying characteristics of the Quicksilver. He did not force or hammer the student beyond their capabilities. Rather, he let the learning experience happen in a natural way. Yet, he was quietly but decisively there if an emergency should occur.

Time seemed to remain still while up in the sky. I was concentrating on flying the Quicksilver and absorbing her individual performance levels. We flew up and down the Western shore of the lake while Chuck drilled me in banks and turns at angles of 20° to 30°, port and starboard. He worked me up into stalls and recoveries. His final instruction on the west side of Lake Elsinore was to take her back to the airpark. But before we landed, he wanted me to do practice landing patterns from a safe altitude of 500 feet AGL.

I slid the Quick into a 30° bank and rolled her around towards the Southern shore about a half mile distance from where the DC-2 sat. I spoke. "I notice you never fly straight across the lake." Chuck looked at me. "I don't like the idea of landing a plane with wheels in the water. I prefer land. Fishing an ultralight out of the lake is not much fun. I've seen a few of them in the water here."

His words made sense. Why take a chance when you don't have to. Fly the sensible and safe route when you have a choice. One of the lessons.

Taking the Quick back to the air park I had a chance to settle in and relate to the plane and the flight. I scanned the wings and flying cables. The fabric was drumhead taut as were the cables. The vibrations from the engine set a buzz throughout every ounce of the Quick and her cargo. She was finely tuned.

My eyes focused on landmarks I had located that were targets for commencement of airport traffic moves. A rusted car, half-submerged in the lake was about one hundred feet from the end of the runway. I use the wreck as my turning point, either going in or flying out. I set course for the wreck in the water.

Just before we reached my mark Chuck asked me to make a heading down an inlet of water between the runway and a finger of land. "Fly over the inlet and use it for the downwind leg of the pattern."

Before following Chuck's order I scanned the entire area for other activity in the air around us and seeing it was clear I set a course to enter the pattern. A few moments later I inserted the Quick into the Downwind Leg of the Airport Traffic Pattern at a 45° angle about 500 feet AGL before we reached the wrecked car. A few moments later we were heading right over my mark. I retarded the throttle to 1/4 thrust, and fine tuned my consciousness on the Quick's heading. I was not holding the stick loosely. I was tense as I set the glide angle.

The engine stopped. Dead. We were 500 feet AGL over the lake.

I pulled back on the stick with a jerk as Chuck's hand slammed against my wrist and shoved my hand and the stick hard forward. We pitched into a dive. He held my hand there as he yanked the starter pull handle down and the Rotax rattled to life. He ran up the throttle, rotated back on the stick and leveled us out a few quiet, thoughtful moments later.

Chuck. "Rick, what was your first reaction when the engine quit?"

"I pulled back on the stick."

"Why did you pull back on the stick?"

"It was an automatic reaction. I pulled back."

"What was your reason for pulling back on the stick?"

I spoke my thought. "I pulled back to hold altitude. My mind was not prepared to make a response. I reacted instinctively out of fear of crashing."

"Pulling back on the stick was the wrong thing to do. Now answer me this. Why did I push the stick forward?"

I set my eyes on the water storage tank that sat on the top of a hill as we passed over it. It also was one of my aerial targets for maneuvering. Chuck's question swirled in my head. I answered with what was on my mind.

"At the time you did it I did not know why I pulled back. You caught me completely by surprise."

"Do you know now why I pushed your hand forward?"

"Yes."

"Why?"

"To gain airspeed."

133

"That is right. Now. Why is airspeed important when you are in the landing pattern?"

The answer boomed into my mind clearly.

"Control. You pitched us forward and put us into a steep dive to maintain stability and control."

"You are right. Airspeed does mean control and when you are landing or taking off, you need all the control you can get. You want the airplane to go where you want it to go, not where it wants to go. When you pulled back on the stick you pitched the nose up into the air and in doing that you increased the angle of attack of the wing to the point where we would have gone into a stall within moments. You committed a serious error. Many pilots have suffered fatal accidents because of that mistake you made. You do not want to loose control when landing or taking off. Your objective is to gain control. Do you understand?"

I banked around the landmark water tower and made a heading back towards the insertion point in the downwind leg as I answered.

"Now I understand. Very clearly."

Chuck continued. "What you did Rick is what just about everyone does. Pull back when the engine stops. Now you know that to blindly follow your instincts can lead you into serious trouble. When you know aerodynamics by heart and know what to do in emergencies you are following facts, not instincts. That is the instruction for survival in flying emergencies. Fly the airplane and maintain airspeed at all times."

I glanced at Chuck as I spoke. "This is a lesson I will not forget."

Chuck looked at me. "I hope you won't forget it. Many have and they have paid for it. Now let's get back into the pattern and do some aerial takeoffs and landings."

We went around the pattern several more times while I practiced approaches, banks, glide slopes, roundouts, flares, engine up, climb outs, level flight then back through the exact routine several times. When Chuck thought I had had enough for the day, he took the controls and brought the Quicksilver in. He held his hand over mine on the stick while we made the landing.

It had been a good day's lesson. I had a lot to think about.

# DOWNEY CA POLICE DEPARTMENT
## QUICKSILVER POLICE INTERCEPTOR

**SEVERAL POLICE DEPARTMENTS** are using ultralights for law enforcement. These ultralights are all special built aircraft and have FAA registered "N" numbers. The police officers who fly them are excellent pilots and can land and take off from difficult spots. The slow speeds and low operational costs of the ultralight allow the police to fly over areas that helicopters or other aircraft are incapable of.

**ROTATION**

**TAKEOFF**

**CLIMB-OUT**

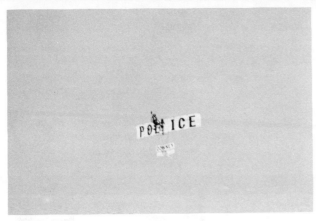

**MANEUVERS**

**FINAL** — Note the telephone wires in the final approach to the parking lot.

**ROUND-OUT**

**TOUCHDOWN**

**LAKE ELSINORE** as seen from the top of the hills that lay to the West of the lake. On the Southeastern end of this lake is the Ultralight Flight Park where I took flight training on the Quicksilver MX. Many times I sat here and reviewed the events of each day of my training. It is an inspirational spot and view. Horus, my Great Dane joined me there on a return visit.

# DAY 3-4 QUICKSILVER

The next two days of the training were spent refining the essentials from preflight, takeoff, climbout, level maneuvering, climbing, descending, landing approaches, landings, taxi, parking, securing the ultralight and flying skill review sessions after each flight.

I found that taking flight lessons in the two place trainer as opposed to tow training had several distinct advantages. The first was the instructor being right there with you all the time. This fact alone built confidence in me, the student. I knew that should an emergency occur that was beyond my flying experience the instructor would be there to solve the problem. On the other hand, learning to fly while under tow built a great deal of confidence in me, the student, as I had to solve each and every flight problem. With the Eagle XL, my instructor Kris watched my maneuvering and corrected the glitches before they developed into bad habits. However, on the whole, I feel that learning to fly in the two place trainer is the best way for the non-pilot to go. The instructor can and will be able to advise and instruct by demonstration while in the air rather than on the ground. This fact alone validates the two place trainer over the tow training system.

The fourth day of my training finished on Chuck's 26th birthday. I secretly bought him a surprise birthday cake and the lunch gang from Eipper, Bruce, Lucky and the three cop pilots from the Downey police department all had a lot of fun together that afternoon. As we parted Chuck informed me that tomorrow morning I would solo in the Quicksilver MX.

# SOLO—
# QUICKSILVER MX

Flying is magic. With eyes as keen as an eagle you scan the earth below. An intense alert concentration focuses on the sights and smell of a never before felt freedom.

Your first bath in the air alone two thousand feet above the ground tingles every nerve and sense in body and mind.

Above you the clouds. Below you the land glides slowly by, a palette of rich earth colors.

You calm your body with a deep breath as you take it all in. You've done it. You are up here in the sky by your own choice and skill. Rushes of pleasure and apprehension shimmer through each muscle fiber as you study the land beneath you. Thoughts and glances fix on the ultralight that supports you.

The left hand cautiously moves from the throttle. Fingers press and clutch the steel rigging. You recall how loose they were on the ground. Plenty of slack. But up here where the birds fly they are as taut as a tuned bass guitar string.

Hands back on the throttle. A slight adjustment and she flies like a dream.

You deserve the reward. The cheers are heard in the wind singing in the cables as the propeller roars thunderously on. A gentle slow pull back on the throttle lowers the engines throat and a steady hum settles down the vibration of the two-cycle brute at full throttle.

You see it all from above. The reward is in your hands but this is no place for thoughts of glory. Eyes sharpen on safe landing targets on the hard land winding away behind. Second rushes chill the dampness in the center of your back as a gentle move of the control stick to the right banks the beautiful airplane

into a smooth three hundred and sixty degree turn.

Hold it a little longer as a steady force on the stick tries to return to neutral.

"Keep the nose up." Words driven into my brain by my instructor. A little more power. Just a touch forward on the throttle and the plane leaps at the horizon. Slowly you allow the stick to return to center.

The plane levels and heads for the mountain tops beyond Lake Elsinore. Gingerly you recall instructions and relax tight fingers on the stick. She flies straight and true, a marvel of flight in the Quicksilver MX.

A glance at the air speed indicator and you hear the instructors last words before takeoff.

"Keep it flying at 45 mph. No less than thirty-five for cruise."

The left hand obeys. With soft tugs on the throttle you inch it back in millimeters until the right hum tunes in for a perfect setting. Another look around with a smile.

Fly the prize up high. Bank and turn, port and starboard. See it all. It's yours. So beautiful. So serene. Life on earth floating by softly there below.

The chill returns. You are totally concerned now with one thing. To be able to fly up here again. To re-experience what you now are feeling. Then, to land safely, so that this thrill is irrevocably confirmed.

*May your weather be calm*
*and your takeoff smooth*
*have an exhilarating flight*
*and for the conclusion of your flight*
*gentle winds for your landing . . .*

# ULTRALIGHT MANUFACTURERS AND AIRCRAFT NAMES

The words "Kit" and "Factory Built" are also included where applicable. Since prices are subjected to constant changes, they are not included. If you are interested in a particular model please write or call the manufacturer for up-to-date details.

CAUTION: It is not recommended by the author to purchase, build, or fly any ultralight without first receiving proper flight training from a certified Flight Instructor. To do otherwise is to encourage disaster for yourself and others.

On inquiries to the manufacturer be certain to ask where their nearest Dealer Training Center is to your specific location.

**COBRA–KING COBRA–HUSKI**
(Kit or Factory Built Through Dealer)
323 N. Ivey Lane
Orlando, FL 32811
305/298-2920

**VECTOR** (Kit)
Aerodyne Systems
194 Millers Falls Rd.
Turners Falls, MA 01376
413/863-9736

**AEROSTAT-340**
(Kit or Factory-Built)
Aerolight Flight Development
456 E. Juanita
Mesa, AZ 05204
602/892-2955

**AVENGER** (Kit or Factory-Built)
Airborne Wing Design
7572 Telser Way
Sacramento, CA 95823
916/395-3374

**CADET JR.** (Kit)
Aircore Industries, Inc.
3726 13th Ave. SE
Bellevue, WA 98006

**HOVEY DELTA BIRD,**
**HOVEY DELTA HAWK** (Bi-Planes)
**HOVEY WAING DING**
(Plans only)
Aircraft Specialties
P.O. Box 1074
Canyon Country, CA 91250
805/252-4054

**SUN FUN–VJ-24 W** (Kit)
Airway Aircraft, Inc.
905 Air Way
Glendale, CA 91201
213/247-4285
Also at:
Volmer Aircraft
P.O. Box 5222
Glendale, CA 91201
213-247-8718

**EAGLE-B, EAGLE-XL, EAGLE**
**2-PLACE, FALCON**
(Purchase through Dealers)
American Aerolights, Inc.
700 Comanche NE
Albuquerque, NM 87107
505/822-1417

**SADLER VAMPIRE**
American Microflight, Inc.
7654 E. Acoma Drive
Scottsdale, AZ 85260
602/951-9772

**CHICKEN HAWK** (Kit)
Atlantis Aviation
4230 Hoff Rd.
Bellingham, WA 98225
206/733-4986

**MITCHELL WING-B-10,**
**MITCHELL WING-A-10,**
**MITCHELL WING-TU-10 (2 Place)**
Aviation Marketing Int'l Ltd.
1870 Wildcat Dr.
Porterville, CA 93257
209/781-2475

**SAND PIPER** (Plans, Kits)
B & B Aircraft Co.
2201 E. Second St.
Newberg, OR 97123
503/538-8855

**SPARROW-G**
B & G Aircraft Co.
115 S. Prospect St.
Hartville, OH 44632
216/877-9909

**MACH-07** (Plans)
Beaujon Ultralights-UA
P.O. Box 2121
Ardmore, OK 73401

**CHINOOK** (Kit)
Birdman Enterprises, Ltd.
7939 Argyll Rd.
Edmonton, Alberta
T6C 4A9, Canada
403/466-5360, 466-2579

**KASPERWING-180-B,**
**KASPERWING-180-BX**
(Kit and Factory-Built)
Cascade Ultralites, Inc.
Issaquah, WA 98109
206/392-0388

**CGS HAWK** (Kit)
CGS Aviation, Inc.
1305 Lloyd Rd.
Wickliffe, OH 44092
216/943-3064

**JENNY** (Kit)
Cloud Dancer Aeroplane Works
P.O. Box 14202
Columbus, OH 43214
614/548-5456

**CLOUDBUSTER** (Kit)
Cloudbuster Ultralights, Inc.
715 Packinghouse Rd.
Sarasota, FL 33582

**CONDOR-II, III, III+2 (2-Place)** (Kits)
Condor Aircraft
10772 SW 190th St.
Miami, FL 33157
305/238-3920

**RAVEN (2-Place)** (Kit)
Custom Aircraft Conversions
222 W. Turbo Dr.
San Antonio, TX 78215
512/349-6347

**NOMAD-II, HONCHO-II,**
**SUPER HONCHO** (Kits)
Delta Technology
12953 East Garvey Blvd.
Baldwin Park, CA 91706
213/814-1467

**HOVEY DELTA BIRD,**
**HOVEY DELTA HAWK**
(Factory Built)
Deltadyne Mfg.
Star Route, Box 104A
El Mirage Airport
Adelanto, CA 92301
619/388-4273

**SNOOP, SNOOP-II** (Kits)
Eastern Ultralights
P.O. Box 424
Chatsworth, NJ 08019
609/726-1193

**QUICKSILVER-MX; QUICK-**
**SILVER-MXL, QUICKSILVER-MX-II**
**(2-Place), POLICE INTERCEPTOR**
(Kits)
Eipper Aircraft Co.
26531 Ynez Road
Temecula, CA 92390
714/676-3228 714/676-6886

**RK-I** (Kit)
Experimental Aeroplane Works
P.O. Box 457
Byron, CA 94514
209/465-0945

**FP-101, FP-202, KOALA** (Kits)
Fisher Flying Products, Inc.
Rt. 2, Box 282
South Webster, OH 45682
614/778-3185

**HUMMINGBIRD-103,**
**PROSPECTOR** (Kits)
Gemini International Inc.
1080 Linda Way, #2
Sparks, NV 89431
702/331-3638

**WACO II** (Factory-Built)
Golden Age Aircraft Co.
P.O. Box 846
Worthington, OH 43085
614/431-1277

**WITCH (Foldable Wings)**
(Factory-Built)
Greenwood Aircraft Corp.
P.O. Box 401
Alexandria, MN 56308
612/762-2020

**BUCCANEER (Wheels),**
**BUCCANEER (Amphibious)**
(Factory-Built)
HighCraft Corp.
P.O. Box 899
Longwood, FL 32750
305/831-6688

**SUNSEEKER (Wing Folds)**
(Factory-Built)
Hillside Aviation
2600 Gold St.
Redding, CA 96001
916/241-4204

**INDIANA FLYER** (Kit)
Industrial Electric
55339 Ash Road
Osceola, IN 46561
219/674-8460

**KOLB FLYER, ULTRA STAR** (Kits)
Kolb Co.
R.D. 3, Box 38
Phoenixville, PA 19640
215/948-4136

**LAUGHING GULL** (Factory Built)
Laughing Gull Aircraft, Inc.
819 S. Vulcan #1
Encinitas, CA 92024
619/942-9563

**PINTAIL, PINTAIL TRAINER**
**(2-Place)** (Kits)
The Little Airplane Company
P.O. Box 255843
Sacramento, CA 95865
916/424-2413

**MAVERICK** (Kit)
Maverick Mfg., Inc.
12139 Glenwood Rd. S.W.
Port Orchard, WA 98366
206/876-9175

**DRIFTER, DRIFTER-XP,**
**HUMMER** (Kits)
Maxair Sports, Inc.
32 Water St.
Glan Rock, PA 17327
717/235-2107

**MEADOWLARK-C** (Factory-Built)
Meadowlark Ultralight Corp.
P.O. Box 1524
Medford, OR 97501
503/779-8284

**TOMCAT TOURER**
(Kit or Factory-Built)
Midwest Microlites
1351 W. Second St.
Oconomowoc, WI 53066
414/567-6663

**MIRAGE II** (Kit)
Mirage Aircraft, Inc.
31 Pearson Way
West Springfield, MA 01089
413/732-5067

**NORTHSTAR-VIKING** (Kit)
Northstar Ultralights, Ltd.
5221 W. Montebello #15
Glendale, AZ 85301
602/931-9462

**PARAPLANE (Flying Parachute)**
(Kit)
ParaPlane Corp.
5801 Magnolia Ave.
Pennsauken, NJ 08109
609/663-2234

**P-CRAFT** (Kit)
Paup Aircraft, Inc.
Arthur Neu Airport
Carroll, IA 51401
712/792-5816

**PETIT BREEZY ULTRALIGHT**
(Plans Only)
Petit Breezy Ultralight
2450 No. 56th St.
Mesa, AZ 85205

**FLIGHT STAR, DUAL STAR**
**(2-Place)** (Kits, Factory-Built)
Pioner International Aircraft, Inc.
P.O. Box 631
Manchester, CT 06040
203/644-1581

**ASCENDER II, ASCENDER II +,**
**ASCENDER II +2,**
**FLEDGLING, LIGHT FLYER,**
**PTRAVELER, PTIGER** (Kits)
Pterodactyl Ltd.
Box 191
Watsonville, CA 95076
408/724-2233

**CHALLENGER** (Kit)
Quad City Ultralights
3610 Coaltown Road
Moline, IL 61265
309/764-3515

**COYOTE** (Kit)
Rans
1104 E. Hwy, 40 Bypass
Hays, KS 67601
913/625-6348

**RITZ STANDARD-MODEL A**
*(Plans, Kit)*
Ritz Aircraft Co.
Shipmans Creek Rd.
Wartrace, TN
615/857-3419

**B1-RD, B2-RD (2-Place)** *(Kits)*
Robertson Aircraft Corp.
Snohomish County Airport
Everett, WA 98204
206/355-8700

**RALLY SPORT, RALLY 2-B,
RALLY-3 (2-Piece)** *(Kits)*
Rotec Engineering, Inc.
Box 220
Duncanville, TX 75116
214/298-2505

**EXCELSIOR** *(Kit)*
St. Croix Ultralights
5957 Seville St.
Lake Oswego, OR 97034
503/636-4153

**LITTLE B1** *(Plans, Kits)*
Simpson Midwest Ultralights
Rt. 1, Box 114
Fisk, MO 63940
314/686-3578

**HIPERLIGHT** *(Kit, Factory-Built)*
Sorrell Aircraft, Ltd.
16525 Tilley Rd. So.
Tenino, WA 98589
206/264-2866

**HORNET** *(Factory-Built)*
SR-1 Enterprises
2323 Endicott
St. Paul, MN 55114
612/646-3884

## PUBLICATIONS
### PILOT TRAINING BOOKS

Federal Aviation Administration
Airman's Information Manual
U.S. Department of Transportation
800 Independence Ave., S.W.
Washington, D.C. 20591

A.S.A. Publications
Pilot Training Program
7201 Perimeter Rd. South
Seattle, WA 98108
206-763-0277

Sky Master International
Ultralight Training System
Pilot Training Program
5205 S. College Ave., Suite B
Fort Collins, CO 80525
303-223-7500

Aviation Book Company
1640 Victory Blvd.
Glendale, CA 91201
213-240-1771

Ultralight Publications
Best Selection—Free Catalog
Pilot Training & Know-How,
Equipment
One Ultralight Way, Box 234
Hummelstown, PA 17036
717-566-0468

Powered Ultralights & Training
Dennis Pagen
P.O. Box 601
State College, PA 16801

Flight Patterns
Ultralight Handbook
26 E. University St.
Tempie, AZ 85281
602/968-5518

**STARFIRE, TRISTAR, TX-1000**
*(Plans & Kits)*
Starflight Mfg.
Rt. 3, Box 197
Liberty, MO 64068
816/781-2250

**SKY WALKER** *(Kit)*
Sterner Aircraft
P.O. Box 811
Sterling Heights, MI 48078
313/268-1882

**STRATOS, STRATOS II (2-Place)** *(Kits)*
Stratos Aviation
21200 Superior St.
Chatsworth, CA 91311
213/882-5500

**LONE RANGER SILVER CLOUD,
SKY RANGER, SILVER CLOUD
(2-Place)**
*(Kits, Factory-Built)*
Striplin Aircraft Corp.
Box 2001
Lancaster, CA 93439
805/945-2522

**MONO-FLY**
*(Plans, Kit, Factory-Built)*
Teman Aircraft Inc.
P.O. Box 1489
Hawaiian Gardens, CA 90716
213/402-6059

**TERATORN, TA-TERA II (2-Place)**
*(Kits)*
Teratorn Aircraft
1604 South Shore Dr.
Clear Lake, IA 50428
515/357-7160

**PEGASUS-II, PEGASUS SUPRA**
*(Factory-Built)*
TFM, Inc.
705 E. Gardena Blvd.
Gardena, CA 90248
213/532-2030

Hawk Publications
Ultralight Experience
46 Sellars Place N.W.
Fort Walton Beach, FL 32548
904-243-8802

### DIRECTORY

International Ultralight Aviation
Directory
70 Kairistine Lane
Winnipeg, MB-R2r-1E8, Ph.
204/633-1373

### PERIODICALS

Glider Rider
P.O. Box 6009
Chattanooga, TN 37401
615/867-4970

Ultralight Flyer
P.O. Box 98786
Tacoma, WA 98499
206/588-1743

Ultralight Aircraft
16200 Ventura Blvd., Suite 201
Encino, CA 91436
800/321-3333

Air Progress Ultralights
10968 Via-Frontera
San Diego, CA 92127
619/485-6535

Ultralight and Sport Aviation
EAA Ultralight Assn.
11311 W. Forest Home Ave.
Franklin, WI 53130

## ENGINES
**CUYUNA**
Cuyuna Engine Co.
P.O. Box 116
Crosby, NM 56441
218/546-8313

**AEROPLANE-XP** *(Kit)*
UFM of Kentucky
2700 Freys Hill Road
Louisville, KY 40222
502/245-0779

**ULTAVIA** *(Kit, Factory-Built)*
Ultavia Aircraft, Inc.
P.O. Box 3316
Las Vegas, NV 87701
505/425-6054

**INVADER** *(Plans, Kit)*
Ultra Efficient Products, Inc.
1637 7th St.
Sarasota, FL 33577
813/955-0710

**LAZAIR, LAZAIR II (2-Plane)**
*(Kits)*
Ultraflight Sales, Ltd.
P.O. Box 370, Port Colbourne, Ont.
L3K 1B7 Canada
416/735-8352

**PHANTOM** *(Kit)*
Ultralight Flight, Inc.
480 Hayden Station Rd.
Windsor, CT 06095
203/683-2760

**EASY RISER** *(Kit—No Engine-Gear)*
Ultralight Flying Machines
P.O. Box 2967
Turlock, CA 95381
209/634-6134

**APOLL II (2-Place)** *(Kit)*
Ultralight Marketing International
111 Elm Street, #203
San Diego, CA 92101
619/296-7722

**WIZARD-J-2, WIZARD-J-3,
WIZARD-T-38 (2-Place)** *(Kits)*
Ultralite Soaring, Inc.
3411 N.E. 6th Terr.
Pompano Beach, FL 33064
305/785-7853

ROTAX
Pacific Flight Engineering
P.O. Box 652
Salinas, CA 93902
408/422-5226

KAWASAKI
Advance Engine Design, Inc.
P.O. Box 589
Flint, MI 48501
313/742-0602

ARROW COMPANY, INC.
P.O. Box 708
Meeker, OK 74855
415-729-3833

## HELMETS
Gentex Corp.
2824 Metropolitan Place
Pomona, CA 91767
714/596-6512

Land Tool Co.
650 E. Gilbert St.
Wichita, KA 67211
316/265-5665

Monarch Helmets
3648 Main St.
Chula Vista, CA 92011
800/854-2096

Wills Wing, Inc.
1208-H.E. Walnut
Santa Ana, CA 92701
714/547-1344

Simpson Safety Helmets
22630 S. Normandie Ave.
Gardina, CA 90502
213/320-7231

## INSURANCE
Lighting Insurance
P.O. Box 16
Westerville, OH 43081
614/882-5135

**LE PELICAN** *(Kit)*
Ultravia, Inc.
795 L'Assomption
Repentigny, Quebec
J6A 5H5 Canada

**SUN FUN VJ-24W** *(Plans)*
Volmer Aircraft
P.O. Box 5222
Glendale, CA 91201
213/247-8718

**WEEDHOPPER** *(Kit)*
Weedhopper of Utah, Inc.
1148 Century Drive, Box 2253
Ogden, UT 84404
801/621-3941

**BARNSTORMER, WOODHOPPER,
GOLD WING, MOHAWK,
BOOMERANG, MITCHELL-B-10,
SKY PUP** *(Kits)*
Wicks Aircraft Supply
410 Pine St.
Highland, IL 62249
618/654-7447

**SKY RAIDER, SKY RAISER-SS,
SPITFIRE** *(Kit)*
Worldwide Ultralite Industries
27711 Interstate 10
Katy, TX 77450
713/392-7000

**WREN** *(Kit, Factory-Built)*
Wren Aviation, Inc.
6315 S. Hydraulic
Wichita, KS 67216

**ZIPPER** *(Factory-Built)*
Zenair / Atlanta
Route 12, Box 720
Gainesville, GA 30501
Zenair / Seattle
606 S.W. 302nd
Federal Way, WA 98003

Stephen P Heekin Agency
1007 Enquirer Bldg.
Cincinnati, OH 45202

## PARACHUTES
Ballistic Recovery System
9242 Hudson Blvd.
Lake Elmo, NM 55042
612/731-1311

Parachute Associates
P.O. Box 811, 145 Ocean Ave.
Lakewood, NJ 08701
201/367-7773

Sky Master Parachute
Para-Gear Equipment Co.
3839 W. Oakton Ave.,
Skokie, IL 60043
312/679-5905

Delta Wing Kites & Gliders
P.O. Box 483,
Van Nuys, CA 91408
213/787-6600

Wills Wing, Inc.
1208 H.E. Walnut
Santa Ana, CA 92701
714/547-1344

G.Q. Security Parachutes, Inc.
P.O. Box 3069-A
San Leandro, CA 94578
415/357-4730

Midwest Parachutes
22799 Heslip Drive
Novi, MI 48050
313/349-2105

Second Chantz, Inc.
P.O. Box 1817
Crystal Bay, NV 89402

# A FINAL THOUGHT

When you lift off into the sky and fly, the other eagles up there welcome you. Their primary concern is whether you are a friendly or an attack eagle.

To remain in the sky for many flights you must be a friendly eagle for the badly trained, non-thinking attack birds are tragically short-lived.

Just as an eagle fledgling is taught to fly by its parents—you must acquire your flying skills from the experts. To do otherwise can lead you too close to the attack eagle status.

Choosing the right flight instructor is your most important assignment. Preflight your intended flight instructor as carefully as the airplane and make certain the CFI (certified flight instructor) appears before his or her name. This first choice of yours is the most important one you must make. Accept nothing less than the best.

I end with an excerpt from an article by a pilot member of the U.S. Aerobatic Team, Gene Beggs of Midland, TX. The article appeared in *Sport Aerobatics* and was reprinted in the August 1984 EAA *Sport Aviation* magazine.

## THE END (OF CONFUSION AND MYSTERY ABOUT SPINS!

### FOR EMERGENCY SPIN RECOVERY

1. **Cut the throttle!**
2. **Take your hand off the stick!**
3. **Kick full opposite rudder until the spin stops!**
4. **Neutralize rudder and pull out of the dive!**

No matter who you are or what your level of experience is, please go back and read it again! If you fly aerobatics, or if you fly an airplane that is capable of spinning, you should know this life saving method of spin recovery. This method of recovery will enable you to quickly and easily recover from **any** spin that can be encountered in any of the airplanes that I have used in the spin tests conducted during the past two years. This method has many advantages over those shown in most aircraft manuals. It is as simple as one, two, three and can be relied on in an emergency situation where a pilot may not be thinking clearly. **It has the added advantage of it being unnecessary for the pilot to know what kind of spin he is in, the recovery procedures are the same whether the spin is upright or inverted, flat or normal, power on or off or otherwise.**

*You can learn to fly an ultralight in one week but it takes continuous refinement of your flying skills to eliminate those underlying human factors that lead to pilot errors.*

### U.S.A. STATISTICS 1982-83

| | |
|---|---:|
| POPULATION Estimated—4/10/84 | 235,353,467 |
| DRIVERS LICENSE—ACTIVE | 150,000,000 |
| LICENSED—Autos—Trucks—Busses | 163,000,000 |
| MOTORCYCLES—Registered—(Est.) | 6,021,000 |
| AIRMEN—Certified—1982 | 764,182 |
| AIRMEN—Certified—1983 | 718,004 |
| AIRCRAFT Shipped—1981—General Aviation | 10,114 |
| AIRCRAFT Shipped—1983—General Aviation | 3,586 |
| HOMEBUILT AIRCRAFT—Experimental—Certified | 10,000 |
| HOMEBUILT AIRCRAFT—Under Construction | 20,000 |
| ULTRALIGHTS Shipped—1978-1982 | 36,501 |
| ULTRALIGHTS Shipped—1983 | 9,989 |

Source: F.A.A., U.S. Census Bureau, A.A.O.P.A. E.E.A.